Animal feeding and food safety

FAO
AND
ION
PER

69

Report of an FAO Expert Consultation
Rome, 10-14 March 1997

Food
and
Agriculture
Organization
of
the
United
Nations

Rome, 1998

Reprinted 1999

M-82
ISBN 92-5-104158-X

CONTENTS

INTRODUCTION

An FAO Expert Consultation on Animal Feeding and Food Safety was held at FAO Headquarters in Rome from 10 to 14 March 1997. The Consultation participants are listed in Annex 1. The Consultation was opened by Mr. John Lupien, Director of FAO's Food and Nutrition Division, who welcomed the participants on behalf of the Director-General of FAO, Dr. Jacques Diouf.

In welcoming the participants, Mr. Lupien pointed out that FAO has had a long-standing interest in the relationship between animal feeding, trade and food safety[1]. Over the years problems such as *salmonellae* and other pathogenic micro-organisms in feed; aflatoxin contamination in feed affecting poultry and trout and other mycotoxin problems; contamination of feeds with pesticide residues, heavy metals, and industrial chemicals have created concern at national and international levels. Such problems can pose risks to human health and significant difficulties to trade in feed and in food derived from animals. In the past many feed components have been handled in ways that were not designed to assure the quality and safety of the final feed. Most recently the link that has been drawn between bovine spongiform encephalopathy (BSE) and feed ingredients has given additional impetus towards devising and enforcing strict quality and safety control procedures in all steps of producing, processing and utilization of feeds.

Mr. Lupien said that the report of this Consultation will help FAO further develop, at the international level, the overall scientific basis that is essential to the development of improved practices in the feeding of animals for the production of food. A recommended code of practice for good animal feeding would improve overall feeding practices and ensure better quality and safer feeds, and better quality and safer animal products for human consumption. The report of this Consultation will be of vital interest to FAO, its member governments and the Codex Alimentarius Commission.

The Consultation elected Professor David Fraser as Chairman and Dr. John Wilesmith as Vice-Chairman. Dr. Keith Behnke was appointed as Rapporteur. In his opening remarks Professor Fraser pointed out that because the cost of feeding was the major expense in many animal production systems, the animal industries are constantly seeking novel and cheaper feeds, all of which may or may not introduce new contaminants into the food chain. He expressed the hope that the Consultation would lead to the formulation of a code of practice for good animal feeding which would minimise hazards which might arise from feeds during livestock production.

BACKGROUND

Food production is a complex process, with the ultimate objective of the food industry and food safety regulators being to ensure that food reaching the consumer is safe and wholesome. Food generally expected to be safe may become unsafe due to the introduction of hazards during production, processing, storage, transport, or final preparation for consumption. For food derived

---

[1] Throughout this report, 'food' means any substance, whether processed, semi-processed or raw, of animal origin, which is intended for human consumption, and includes milk. The word 'feed' means any substance whether processed, semi-processed or raw, which is intended for consumption by animals from which food is derived.

from animals, the hazard may originate from a number of these and other sources including the consumption by food production animals of contaminated feed.

Examples of hazards in food that can be linked to feed and have long been recognised include salmonellosis, mycotoxicosis, and the ingestion of unacceptable residue levels of veterinary drugs and agricultural and industrial chemicals. In addition, if the postulated link between BSE and the new variant of Creutzfeldt-Jakob Disease (nv-CJD) is established, it would be another example of food contamination originating in feed.

Two papers were commissioned by FAO for the Consultation. The first paper (Annex 3) addresses feed processing and the second paper (Annex 4) addresses infections and intoxications of farm livestock associated with feed and forage.

The Joint FAO/WHO Food Standards Programme, administered by the Codex Alimentarius Commission (CAC), has also done some work in this area in the past. Included in its standards, guidelines and other recommendations are quality and safety standards for meat and various meat products, maximum levels for contaminants, maximum residue levels for residues of veterinary drugs and pesticides and codes of practice ranging from hygienic practices to use and control of veterinary drugs. Annex 5 contains a description and summary of the Codex work.

SCOPE

The Consultation restricted its considerations to food safety matters that pertained strictly to feeds. It did not consider plant toxins or radionuclides, nor did it consider parasites such as *Taenia saginata* that are spread by human sewage. In addition, the risks to human health from feed or forage contaminated with several other agents such as *Bacillus anthracis, Clostridium botulinum* toxin, *Listeria* spp., *Mycobacterium bovis* and *Yersinia* spp. appear to be negligible or non-existent and were therefore not considered by the Consultation. It also did not consider management practices unrelated to feeding, such as vaccination or other veterinary treatments including the use of injectable agents or drenches. It did not consider spoilage of food products nor did it consider normal feeding practices aimed at maintaining good nutritional status of production animals.

While there are a great many foods that are of animal origin, the Consultation restricted its consideration to those foods from common domestic animals which have significance in international trade. These include meat and meat products, milk and milk products, eggs and egg products, and products of aquatic animals derived from aquaculture. All animal feeds were considered other than natural unrestricted grazing. The Consultation limited its considerations to food that complies with CAC recommendations, for example meat judged to be fit for human consumption in accordance with the Recommended International Code for Ante-mortem and Post-mortem Inspection of Slaughter Animals and for Ante-mortem and Post-mortem Judgement of Slaughter Animals and Meat (CAC/RCP 41-1993). As one example, this excluded consideration of foodborne anthrax.

Aquaculture products are a major source of food protein in developing countries and one of the fastest growing systems of food production. There are two broad categories of aquatic animal production. One involves the production of carnivorous/omnivorous fish using intensive farming systems and is largely dependent upon the use of compounded feed, while the other is based on the mass production of herbivorous/filter feeding fish species within semi-intensive/extensive farming systems based on the use of agricultural and other by-products, including animal manure, as fertilizer or supplementary feed inputs. Intensive fish farming using compounded feed therefore clearly fell within the scope of the Consultation. However, as semi-intensive/extensive systems are being considered by the WHO/FAO/NACA[1] Study Group on Food Safety Issues associated with Products from Aquaculture, this was not considered.

INTERNATIONAL TRADE

The Uruguay Round of Multilateral Trade Negotiations established a new World Trade Organization (WTO) and included negotiations on reducing non-tariff barriers to international trade. Included in the Final Act were the Agreements on the Application of Sanitary and Phytosanitary Measures (the SPS Agreement) and on Technical Barriers to Trade (the TBT Agreement). Both Agreements have implications for the work of the Codex Alimentarius Commission.

The SPS Agreement confirms the right of WTO member countries to apply measures necessary to protect human, animal and plant life and health provided that "such measures are not applied in a manner which would constitute a means of arbitrary or unjustifiable discrimination between countries where the same conditions prevail, or a disguised restriction on international trade"[2].

With respect to food safety, the SPS references the standards, guidelines and recommendations established by the CAC relating to food additives, residues of veterinary drugs and pesticides, contaminants, methods of sampling and analysis, and codes and guidelines of hygienic practice.

Therefore, measures need to be taken with respect to feeds to ensure adherence in food of animal origin to the Codex maximum levels or guideline levels for contaminants, and to the Codex maximum residue limits (MRLs) for pesticide and veterinary drugs. Measures also need to be taken to ensure that appropriate hygienic practices are followed at all stages of the animal feeding chain to prevent, eliminate or reduce potential hazards in the food.

The objective of the TBT Agreement is to prevent the use of national or regional technical requirements, or standards in general, as unjustified technical barriers to trade. It covers all types of

---

[1] NACA is the Network of Aquaculture Centres in the Asia-Pacific Region.

[2] The SPS Agreement defines SPS Measures as those measures applied:
- to protect animal or plant life or health within [a country's] territory .... from risks arising from the entry, establishment or spread of pests, diseases, disease-carrying organisms or disease-causing organisms;
- to protect human or animal life or health within the territory of [a country] from risks arising from additives, contaminants, toxins or disease-causing organisms in foods, beverages or feedstuffs;
- to protect human life or health within the territory of [a country] from risks arising from diseases carried by animals, plants or products thereof, or from the entry, establishment or spread of pests; or
- to prevent or limit other damage within the territory of [a country] from the entry, establishment or spread of pests.

standards, including all aspects of food standards other than those related to SPS measures, and includes a very large number of measures designed to protect the consumer against deception and economic fraud. The aspects of food standards it covers relate specifically to quality provisions, nutritional requirements, labelling and methods of analysis. The TBT Agreement basically provides that all technical requirements and regulations must have a legitimate purpose and that the impact or cost of implementing the measure must be proportional to the purpose of the measure. It also places emphasis on international standards.

The Consultation recognised that increased scientific, legal and political demands are being made on the standards, guidelines and recommendations elaborated by Codex. This is in part due to increased consumer interest in food safety, the WTO's SPS and TBT Agreements, harmonization initiatives, calls for increased scientific rigour, the need for transparency, and shrinking national regulatory resources. Therefore, a code of practice for good animal feeding was drafted by the Consultation which would facilitate international trade in animal feedstuffs and animal food products.

POTENTIAL HAZARDS ASSOCIATED WITH FEED[1]

Mycotoxins

Mycotoxins are secondary metabolites produced by fungi of various genera when they grow on agricultural products before or after harvest or during transportation or storage. Some fungi such as *Fusarium* spp. typically infest grains before harvest, others such as *Penicillium* spp. can invade grain after harvest, while *Aspergillus* spp. can grow on grains both before and after harvest. It must be emphasised that the presence of the fungi does not necessarily imply that mycotoxins can be found. Conversely, the absence of fungi does not necessarily mean the absence of mycotoxins.

Both intrinsic and extrinsic factors influence fungal growth and mycotoxin production on a given substrate. The intrinsic factors include water activity, pH, and redox potential whereas extrinsic factors which influence mycotoxin production are relative humidity, temperature and availability of oxygen.

Many mycotoxins, with different chemical structures and biological activities, have been identified. Mycotoxins may be carcinogenic (e.g. aflatoxin B_1, ochratoxin A, fumonisin B_1), oestrogenic (zearalenone and I and J zearalenols), neurotoxic (fumonisin B_1), nephrotoxic (ochratoxins, citrinin, oosporeine), dermonecrotic (trichothecenes) or immunosuppressive (aflatoxin B_1, ochratoxin A, and T-2 toxin). Much of the published information on toxicity concerns studies in experimental animals and these may not reflect their effects in humans and other animals. In addition, the significance of the presence of most mycotoxins in foods of animal origin is not completely understood.

Mycotoxins are regularly found in feed ingredients such as maize, sorghum grain, barley, wheat, rice meal, cottonseed meal, groundnuts and other legumes. Most are relatively stable

---

[1]   The order of appearance of categories of hazards listed (mycotoxins etc.) is not intended to indicate any ranking of relative importance.

compounds and are not destroyed by processing of feed and may even be concentrated in screenings.

Different animal species metabolise mycotoxins in different ways. For example in pigs, ochratoxin A can undergo entero-hepatic circulation and is eliminated very slowly while it is rapidly excreted by poultry species. The polar mycotoxins, such as fumonisins, tend to be excreted rapidly.

Mycotoxins, or their metabolites, can be detected in meat, visceral organs, milk and eggs. Their concentration in food is usually considerably lower than the levels present in the feed consumed by the animals and unlikely to cause acute intoxications in humans. However residues of carcinogenic mycotoxins, such as aflatoxin B_1 and M_1, and ochratoxin A, when present in animal products pose a threat to human health, and their levels should be monitored and controlled.

In most instances the principal source of mycotoxins for humans is contaminated cereals and legumes rather than animal products. This means that the exposure to mycotoxins may be greater in developing countries in which cereal grains and legumes form the staple diet and the intake of animal products, including meat, is low.

There is little information available regarding the occurrence of mycotoxin residues in animal products intended for human consumption. Some examples are summarised in Table 1. Examples of maximum levels in force in various countries include 0.05-1 ppb for aflatoxin M_1, 5 ppb for aflatoxin B_1, 25 ppb and 50 ppb ochratoxin A in porcine kidneys and cereals respectively and, depending on the country, 30-1,000 ppb for zearalenone in corn and foods (1). The levels of mycotoxins detected are usually below the maximum levels accepted in most countries.

Table 1. Examples of food of animal origin which may be naturally contaminated with mycotoxins

Mycotoxin	Potential effects on humans	Occurrence	Maximum level reported	Reference
Aflatoxin B_1	Hepatic cancer	Eggs	0.4 ppb	(2)
		Pig liver	0.5 ppb	(3)
		Pig muscle	1.04 ppb	(4)
		Pig kidney	1.02 ppb	(4)
Aflatoxin M_1	†	Cow's milk	0.33 ppb	(5)
Ochratoxin A	Renal damage	Pig liver	98 ppb	(6)
		Kidney	89 ppb	(7)
		Sausages	3.4 ppb	(7)
Zeralenone	Oestrogenic	Pig liver	10 ppb	(8)
		Pig muscle	10 ppb	(8)

† There is insufficient evidence to describe aflatoxin M1 as a human carcinogen although it is a potent carcinogen in rodents.

Infectious Agents

Transmissible spongiform encephalopathies in ruminants

The transmissible spongiform encephalopathies (TSEs) are non-febrile neurological diseases. They have a long incubation period and are ultimately fatal. TSEs are associated with incompletely defined agents currently termed prions which are resistant to normal heat treatments of feed and food. The TSEs recognised in food producing animals are BSE and scrapie. Sheep scrapie has been recognised for over 250 years. BSE was first recognised in the UK during 1986. The BSE infectious agent enters feed primarily through infected tissues (notably the central nervous system and the reticuloendothelial system) rendered under conditions of insufficient heat treatment to reduce the concentration of the infectious agent to an ineffective dose.

In the case of sheep scrapie, infection is naturally maintained by transmission between sheep. It is likely that humans have been exposed to the scrapie agent by eating brain and other tissues from infected sheep although there is no evidence that the occurrence of either CJD or nv-CJD has been associated with scrapie. With respect to BSE, humans can potentially be exposed through consumption of the infected tissues. The occurrence in humans of nv-CJD has raised the possibility of an association with the BSE agent. At present, with the limited number of diagnosed

6

cases, there is no proven link between nv-CJD and the possible transmission of the infective agent from bovine tissue to humans.

Salmonella enterica

There are over 2,000 salmonella serotypes and these can be divided arbitrarily into three unequally sized groups. These include:

1) the species specific serotypes such as *S. dublin* (cattle) and *S. gallinarum* and *S. pullorum* (poultry);

2) the invasive serotypes which may cause septicaemic disease in several animal species (e.g. *S. enteritidis* and *S. typhimurium*); and

3) the non-invasive serotypes which tend not to result in septicaemia. Members of the first group are not recognised as feedborne pathogens.

The third group is by far the largest and may be associated with subclinical infections in farm livestock. Occasionally they can cause disease and are associated with food poisoning in humans. The principal manifestation of human salmonellosis is gastroenteritis. Septicaemia occurs in a proportion of patients. The case mortality rate is low with the young, old or immunocompromised being most susceptible.

Salmonellae are widely distributed in nature, and feed is only one of many sources for farm animals. Feed ingredients, of both animal and plant origin, are frequently contaminated with salmonellae although the most common serotypes isolated are rarely the most prevalent in animals including man. The two most important serotypes associated with human disease, *S. enteritidis* and *S. typhimurium*, are rarely isolated from feed. Feed can be contaminated by contact with raw ingredients after processing.

Toxoplasma gondii

The protozoan *Toxoplasma gondii* is found in cats, and based on serological surveys, in birds, and in domesticated species including sheep, pigs, goats, and horses. The primary source of infection for animals is feeds contaminated with cat faeces.

Cats are an important source of infection for humans, with the handling or consumption of raw meat also being implicated. Pregnant women who become infected may abort or give birth prematurely, and infants often develop central nervous system disorders and ocular disease. Immunocompromised patients are at particular risk.

Trichinella spiralis

Trichinella spiralis is a nematode which parasitises the intestinal tract of mammals, particularly pigs. The larvae encyst in the tissues, particularly the muscles, which act as a source of infection for humans who consume raw or undercooked meat. The clinical manifestations include fever, muscle pain, encephalitis, meningitis, myocarditis and very occasionally, death.

The cysts can be killed by freezing infected carcasses at -18°C for 20 days. They are also heat sensitive and are killed by traditional rendering temperatures. Effective cooking of raw meat and table scraps before feeding to farm animals will eliminate this hazard.

Veterinary drugs and agricultural and other chemicals

Veterinary drugs

Veterinary drugs may be administered in animal feeds. If the concentration used results in foods of animal origin with residues exceeding the established Maximum Residue Limits (MRLs) such as those established by Codex, there may be a potential risk to human health. Codex MRLs should not be exceeded if concentrations used are correct, withholding times are observed and Good Agricultural Practices (GAP) and Good Veterinary Practices (GVP) are applied.

Agricultural and other chemicals

The potential hazards may include excessive residue levels of herbicides, pesticides, and fungicides and industrial/environmental or other extraneous contaminants such as the polychlorinated biphenyls (PCBs) and heavy metals including mercury, lead, or cadmium. Cereals and treated seeds are the most likely source of these contaminants. The most significant hazards to human health are those chemicals that accumulate in animal tissues or are excreted in milk or become incorporated in eggs at levels in excess of established limits such as the Codex MRLs for pesticides or maximum levels for contaminants in a food or feed.

ASSESSMENT OF THE RISK

The risk analysis approach has been adopted within the Codex system and used as the fundamental method underlying the development of food safety standards. The Consultation therefore saw one of its tasks as to provide an assessment of the risk of foodborne hazards that enter the food chain via feeds.

Risk analysis is composed of three separate but integrated elements, namely risk assessment, risk management and risk communication. Risk assessment has been defined by Codex as being the scientific evaluation of known or potential adverse effects resulting from human exposure to foodborne hazards. The risk assessment process consists of the following steps: (i) hazard identification, (ii) hazard characterization, (iii) exposure assessment, and, (iv) risk characterization. The definition includes quantitative risk assessment, which emphasises reliance on numerical expressions of risk, as well as an indication of the attendant uncertainties. Hazard identification is defined as the identification of known or potential health effects associated with a particular agent. Exposure assessment is the qualitative and/or quantitative evaluation of the degree of intake likely to occur. Risk characterization is the integration of hazard identification, hazard characterization and exposure assessment into an estimation of the adverse effects likely to occur in a given population, including attendant uncertainties. For chemical agents, a dose/response assessment should be performed. For biological or physical agents, a dose-response assessment for hazards should be performed if the data are available or obtainable.

In general terms, the Consultation recognised that there are risks arising from foodborne hazards that enter the food chain via feeds. However, on balance, the judgement was that the risk of these hazards was low in comparison to foodborne hazards that originate from other sources. The risks from *Salmonellae* for example, may be considerably greater during processing of carcasses and subsequent animal product processing. The exposure to mycotoxins is far greater from eating contaminated cereal grains than from eating foods derived from animals fed contaminated grains.

SOURCES OF FEEDBORNE HAZARDS

Feeds can be of animal, plant, microbial or mineral origin. The following tables set out potential hazards and the stage at which they may enter the food chain.

Table 2. Feed of plant origin

	Pre-harvest	Post-harvest	Post-processing
Grains and their by-products	A,B,D	A	A
Oilseeds and their by-products	A,B,D	A,B,D	A,B,C,D
Molasses	B,D	...	A (possible)
Roots and tubers	B,D	A	A
Forages	A,B,C,D	A,C	A,B
Roughages	...	A,B,C	...
Fats and oils	...	...	B

Key: A = mycotoxins, B = agricultural chemicals, C = microbial pathogens, D = metals

Table 3. Feed of animal origin

	Raw materials	Post-processing
Mammalian protein meals	B,C,D,E,F	C
Poultry meals	B,C,D,E	C
Aquatic animal meals	B,C,D	A,C
Fats and oils	B	...

Key: A = mycotoxins, B = agricultural chemicals, C = microbial pathogens, D = metals,

E = drug residues, F = TSEs

Table 4. Miscellaneous feed ingredients

	Raw material	Post-processing
MINERAL ORIGIN		
Phosphate, calcium and sodium sources	D	...
Trace element premixes	D	...
Non-nutritive adsorbents	D	...
MICROBIAL ORIGIN		
Yeast and single cell protein	C,D	D
MISCELLANEOUS ORIGIN		
Food waste	A,C	A,C
Animal manure	B,C,D	C

Key: A = mycotoxins, B = agricultural chemicals, C = microbial pathogens, D = metals

CONTROL OF FEEDBORNE HAZARDS

Feed and feed ingredients should be obtained and preserved in a stable condition so as to prevent hazardous effects due to contamination or deterioration. When received, feeds should be in good condition and meet generally accepted quality standards. Preservation can be facilitated by low temperature storage, ensiling, dehydration or the addition of appropriate chemicals (e.g. propionic acid). Furthermore, pasteurization reduces the numbers of most pathogens. Maintaining low water activity (i.e. A_w <0.65) will minimise bacterial and fungal growth.

Good Manufacturing Practices (GMPs) should be followed at all times. Specific control measures for identified hazards are listed below.

Transmissible spongiform encephalopathies

⇒ All tissues from cattle with clinical BSE should be incinerated so that they are eliminated from all feed and food chains.

⇒ In countries where BSE has occurred, depending on its incidence[1], consideration should be given to placing restrictions on the use of meat and bone meal derived from specific bovine tissues in ruminant feeds. A similar consideration should be made in countries where a risk assessment indicates that the cattle population has been exposed to infection.

---

[1]   As determined by a competent authority using an appropriately structured surveillance programme.

⇒ In countries where BSE and sheep scrapie have occurred, consideration should be given to placing restrictions on the use of ruminant derived protein from the feeds of ruminants.

⇒ In countries where BSE has not occurred, but where sheep scrapie is present, consideration should be given, dependent on the incidence of scrapie and the time/temperature processes used for the rendering of ovine carcasses and tissues, to placing restrictions on the feeding of ovine derived protein to ruminants.

⇒ The measures listed above may require re-evaluation in the light of future research findings on the inactivation of TSE agents during rendering.

⇒ Cross contamination of cattle feeds with meat and bone meal produced from the rendering of potentially infected cattle tissues should be prevented.

Biological agents

⇒ *Salmonella enterica, Toxoplasma gondii, Trichinella spiralis* are sensitive to heat and are readily killed if the manufacture of feed involves pasteurization.

⇒ Protocols developed for GMP must include measures which prevent recontamination of heat treated feed by these agents.

Veterinary drugs

⇒ Only products licensed for administration to food producing animals should be used and the withholding time should be observed before milk or eggs are used for food or animals are sent for slaughter. Adherence to the Codex Code of Practice and Guidelines for Control of the Use of Veterinary Drugs and of Veterinary Drug Residues in Foods (CAC/RCP 38-1993) will ensure animal feeds do not contribute to excessive veterinary drug residues in foods.

Agricultural chemicals

⇒ It is essential that the levels of agricultural chemicals in feed are sufficiently low that their concentration in food is consistently below the established maximum residue limits such as those limits established by Codex.

Mycotoxins

⇒ Feeds contaminated with mycotoxins in excess of established national maximum levels or international established maximum levels such as those established by Codex, should not be fed to animals producing milk, eggs or other tissues used for human consumption.

⇒ Grain and cereals should be stored under conditions of low moisture. Mould inhibitors can be added to reduce fungal growth.

⇒ Mycotoxin contaminated grains can be used for alternative purposes such as alcohol production, but by-products that result should not be fed to food producing animals.

CONCLUSIONS

*   Certain chemical substances and biological agents incorporated into feed, either intentionally or unintentionally, can result in hazards in food of animal origin and may enter feed at any stage of production up to the point of feeding.

*   The risks to human health associated with hazards involved in animal feeding are relatively low in comparison the risks arising from to foodborne hazards from other sources.

*   Where foodborne hazards originate in feed, the hazard should be adequately controlled.

*   Feed ingredients which do not pose any foodborne risk or for which any foodborne risk can be adequately controlled should not be prohibited from use in feed on the basis of food safety concerns.

*   Changes in feed or in the formulation of feed, as well as changes in feed processing methods, may result in changes in the risk from foodborne hazards which originate in feed. It is important that this be recognised and that potential risks be evaluated before any change is made.

*   The management of risk from foodborne hazards which originate in feed needs to be weighed against the potentially greater risks that would result from an inadequate or overly expensive food supply as well as the environmental risk that would result from the failure to recycle nutrients.

*   There is a need for collaboration between all parties involved in feed and animal production, especially those in a position to provide veterinary clinical and epidemiological information, to establish the linkage between any identified or potential hazard and the level of risk. Such information is essential for the development and maintenance of appropriate risk management options and safe feeding practices.

*   Regulatory programmes should be established which ensure that foods of animal origin produced for human consumption are both safe and wholesome. In this context, the hazards which have been identified by the Consultation are well recognised, and suitable and appropriate control measures are in place in many countries. Examples include ante- and post-mortem inspection of slaughter stock, the control of the manufacture and use of veterinary drugs and agricultural chemicals, as well as residue monitoring programmes.

*   Though no conclusive evidence has yet been published, the Consultation determined it to be prudent not to exclude BSE as a potential foodborne hazard. It concluded that the risk that arises from this should be assessed and managed in exactly the same way as other foodborne hazards. This may necessitate the exclusion of certain material from feed for particular circumstances.

*   The disciplines that apply to international trade in both food and feed, as well as in feed ingredients were agreed to during the Uruguay Round of Multilateral Trade Negotiations and set out in the SPS Agreement. The code of practice for good animal feeding that has been developed by the Consultation is intended to provide guidance which will minimise foodborne risks associated with feeds in a manner which is consistent with the principles of the SPS Agreement. The Consultation was of the view that adherence to this code should

12

obviate the need for any trade restrictions on food or feed based on feed related human health concerns.

* With respect to the production of food from aquatic animals using formulated feed, the food safety issues are the same as those applying to the production of food from terrestrial animals and no special issues apply. The Consultation concluded, however, that there may be food safety issues associated with the feeding of aquatic species through the fertilization of ponds by animal manure, agricultural by-products and other wastes. It noted that this issue is to be considered by the WHO/FAO/NACA Study Group on Food Safety Issues on Food Safety Issues associated with Products from Aquaculture and concluded that was the appropriate forum for this issue to be addressed.

RECOMMENDATIONS

The Consultation made the following RECOMMENDATIONS.

1. The feed industry and the animal production industries should recognise their important role in the production of safe food and should evaluate the consequences to human health when using new feed ingredients, new suppliers or introducing new processing methods.

2. As quality assurance is applicable at all stages of food production to ensure the safety of the consumer, a code of practice for good animal feeding should be followed.

3. Manufacturers should provide adequate information to enable the quality and safety of feed to be maintained after delivery.

4. Known and potential risks to food safety should be re-evaluated as new information becomes available.

5. A code of good practice for the fertilization of ponds by the addition of animal manure, agricultural by-products and other wastes should be developed by the WHO/FAO/NACA Study Group on Food Safety Issues on Food Safety Issues associated with Products from Aquaculture and conveyed to the Codex Alimentarius Commission for possible inclusion in a Code of Practice for Good Animal Feeding.

6. The Codex Alimentarius Commission should consider for adoption the Draft Code of Practice for Good Animal Feeding in Annex 2.

7. The feed industry should assist developing countries by providing and promoting advice on good animal feeding practices.

8. FAO should support developing countries in the application of good animal feeding practices.

REFERENCES

1. Van Egmond, H.P., 1989. Current situation on regulations for mycotoxins. Overview of tolerances and status of standard methods of sampling and analysis. *Food Additives and Contaminants*, 6:139-188.

2. Fukal, L. & Sova, Z., 1988. Survey of the presence of aflatoxin in eggs. *Veterinarna Medicina*, 33:675-681.

3. Honstead, J.P., Dreesen, D.W., Stubblefield, R.D. & Shotwell, O.L. 1992. Aflatoxins in swine tissues during drought conditions: an epidemiological study. *Journal of Food Protection*, 55:182-186.

4. Sova, Z., Fibir, J., Reisnerova, H. & Mostechy, J. 1990. Aflatoxin B1 residue in muscle and organs of pigs reared in the pig fattening testing station. *Sbornik Vysoke Skoly Zemedelske v Praze, Fakulta Agronomicka. Rada B, Zivocisna Vyroba*, 52:3-8.

5. Patterson, D.S.P., Glancy, E.M. & Roberts, B.A. 1980. The 'carry over' of aflatoxin M1 into the milk of cows fed rations containing a low concentration of aflatoxin B1. *Food and Cosmetics Toxicology*, 18:35-37.

6. Koller, B. 1992. Occurrence of ochratoxin A in samples of liver and kidney from pigs slaughtered in Steiermark, Austria. *Wiener Tierarztliche Monatsschrift*, 79:1,31.

7. Scheuer, R. 1989. Investigation into the occurrence of ochratoxin A. *Fleischwirtschaft*, 69:1400-1404.

8. Sawinsky, J., Halasz, A., Borbiro, N. & Macsai, G. 1989. Investigation into the mycotoxin content of pork. *Elelmezesi Ipar*, 43:298-299.

LIST OF PARTICIPANTS

EXPERTS

Mr. Errol M. Angeles, Manager, B-MEG Batangas Plant, Luzon Operations Center, San Miguel Foods, Inc., Manila B-MEG, 658 A. Bonifacio Avenue, Balintawak, Quezon City, The Philippines

Dr. J. Stan Bailey, Research Microbiologist, Agricultural Research Service, USDA, Richard B. Russell Research Center, PO Box 5677, Athens GA 30604-5677, USA

Dr. Keith C. Behnke (*Rapporteur*), Dept of Grain Science & Industry, Shellenberger Hall, Kansas State University, Manhattan KS 66506, USA

Dr Mali Boonyaratpalin, Director, Feed Quality Control and Development Division, Department of Fisheries, Ministry of Agriculture and Cooperatives, Kasetsart University, Ladyao, Jatujak, Bangkok 10900, Thailand

Dr. Gonzalo J. Diaz, Associate Professor, College of Veterinary Medicine, National University of Colombia, Apartado Aéreo 76948, Santafé de Bogotá, D.C., Colombia

Professor David R. Fraser (*Chairman*), Dean, Faculty of Veterinary Science, University of Sydney NSW 2006, Australia

Dr. Tata S.P. Hutabarat, Head, Food Safety Sub-Directorate, Directorate of Veterinary Services, Directorate General of Livestock Services, Department of Agriculture, Jl Harsono R.M. No. 3, Ragunan, Pasar Minggu, Jakarta 12550 Indonesia

Dr. Radulf C. Oberthür, Geschäftsführer, Fleischmehlfabrik Brögbern GmbH & Co., Ulanenstr. 1-3, D-49811 Lingen-Brögbern, Germany

Dr. Jos M.A. Snijders, Associate Professor of Meat Hygiene, Department of Science of Food of Animal Origin, Faculty of Veterinary Medicine, Utrecht University, PO Box 80.175, 3508 TD Utrecht, The Netherlands

Dr. Pieter G. Thiel, Specialist Scientist, Programme on Mycotoxins and Experimental Carcinogenesis, Medical Research Council, PO Box 19070, Tygerberg 7505, South Africa

Dr. John W. Wilesmith (*Vice-Chairman*), Head of the Epidemiology Department, Central Veterinary Laboratory (Weybridge), New Haw, Addlestone KT15 3NB, UK

OBSERVERS

Delegate of the Chairman of the Codex Committee on Meat Hygiene and of the Codex Committee on Milk and Milk Products (CCMH & CCMMP)

Dr. Anthony Zohrab, Counsellor (Veterinary Services), New Zealand Mission to the European Communities, Boulevard du Regent 47-48, 1000 Bruxelles, Belgium

Chairman of the Codex Committee on Residues of Veterinary Drugs in Foods (CCRVDF)

Dr. Stephen F. Sundlof, Director, Center for Veterinary Medicine, Food and Drug Administration, HFV-1, MPN-2, 7500 Standish Place, Rockville MD 20855, USA

SECRETARIAT

Mr. Gregory D. Orriss, Chief, Food Quality and Standards Service, FAO, 00100 Rome, Italy *(Secretary)*

Dr. Colin G. Field, Food Quality and Standards Service, Food and Nutrition Division, FAO, 00100 Rome, Italy *(Consultant)*

Dr. Gunter Heinz, Animal Production Service, FAO, 00100 Rome, Italy

Dr. Mike H. Hinton, Division of Food Animal Science, Department of Clinical Veterinary Science, University of Bristol, Langford House BS18 7DU, UK *(Consultant)*

Mr. Q. Dick Stephen-Hassard, PO Box 710, Dillon Montana 59725-0710, USA *(Consultant)*

Dr. Albert Tacon, Inland Water Resources and Aquaculture Service, FAO, 00100 Rome, Italy

DRAFT CODE OF PRACTICE FOR GOOD ANIMAL FEEDING

Introduction

This code of practice applies to feed manufacturing and to the use of all feeds, other than those consumed while grazing free range. The objective of the code is to encourage adherence to Good Manufacturing Practice (GMP) during the procurement, handling, storage, processing (however minimal), and distribution of feed for food producing animals. A further objective is to encourage good feeding practices on the farm.

There are potential risks to human health associated with the contamination of feed with chemical or biological agents. This code outlines the means by which these hazards can be controlled by adopting appropriate processing, handling and monitoring procedures. The principle approaches required for assessing foodborne hazards to human health have been outlined elsewhere.[1*]

General management

The ultimate responsibility for the production of safe and wholesome feed lies with the producer or manufacturer who should produce feeds with as low a level of hazard as possible and comply with any applicable statutory requirements.

The effective implementation of GMP protocols will ensure that:

- buildings and equipment, including processing machinery, will be constructed in a manner which permits ease of operation, maintenance and cleaning;
- staff will be adequately trained and that training is kept up to date;
- records will be maintained concerning source of ingredients, formulations including details and source of all additives, date of manufacture, processing conditions and any date of dispatch, details of any transport and destination;
- water used in feed manufacture is of potable quality;
- machinery coming into contact with feed is dried following any wet cleaning process;
- condensation is minimised;
- sewage, waste and rain water is disposed of in a manner that ensures that equipment, ingredients and feed are not contaminated; and
- feed processing plants, storage facilities and their immediate surroundings are kept clean and free of pests.

---

[*] *Application of Risk Analysis to Food Standards Issues, Report of the Joint FAO/WHO Expert Consultation, Geneva, Switzerland, 13-17 March 1995* (WHO/FNU/FOS/95.3).

Raw materials of animal and plant origin

Raw materials of animal and plant origin should be obtained from reputable sources, preferably with a supplier warranty. Monitoring of ingredients should include inspection and sampling of ingredients for contaminants using risk based protocols. Laboratory testing, where undertaken, should be by standard methods. Ingredients should meet acceptable, and if applicable, statutory standards for levels of pathogens, mycotoxins, herbicides, pesticides and other contaminants which may give rise to human health hazards.

In order to control the spread of specific pathogens it may be necessary to specify, for any given ingredient, the country and species of origin and any treatment process used prior to purchase. Care should be taken to preserve the identity of such material after procurement to facilitate any tracking that might be required.

Minerals, supplements, veterinary drugs and other additives

Minerals, supplements, veterinary drugs and other additives should be obtained from reputable manufacturers who guarantee the concentration and purity of ingredients and provide instructions for correct use.

General management of feeds

Feeds should be stored so as to prevent deterioration and contamination.

Processed feeds should be separated from unprocessed ingredients.

Containers and equipment used for transport, storage, conveying, handling and weighing should be kept clean.

Equipment should be 'flushed' with 'clean' feed material between batches of different formulations to control cross contamination.

Pathogen control procedures, such as pasteurization or the addition of an organic acid to inhibit mould growth, should be used where appropriate and results monitored.

Apart from feeds fed moist, such as silage and by-products of brewing, ingredients and feeds should be kept dry to limit fungal and bacterial growth. This may necessitate ventilation and temperature control.

Waste and unsaleable material should be isolated and identified, and only recovered as feed after freedom from hazardous contamination has been assured. Waste and unsaleable material containing hazardous levels of veterinary drugs, contaminants or any other hazards should be disposed of in an appropriate and, where applicable, statutory manner and not used as feed. If freedom from hazardous contaminants cannot be established, the material should be destroyed.

Packaging materials should be newly manufactured unless known to be free of hazards that might become feedborne.

Labels should be consistent with any statutory requirements and should describe the feed and provide instructions for use.

Feeds should be delivered and used as soon as possible after manufacture.

Personnel

All plant personnel should be adequately trained and should work to GMP standards.

CONTROL OF HEALTH FACTORS
IN THE PRODUCTION OF ANIMAL FEEDS: AN OVERVIEW

SUMMARY

World-wide feed tonnage is of the order of four billion tonnes per annum of which some five hundred and fifty million tonnes is manufactured, or complete feed. FAO has sought independent expert advice in order to identify and determine control measures applicable to human disease transmitted through feed up the food chain to human consumers of animal products. This paper addresses current good manufacturing practices which apply mainly to the manufacture of complete feeds. Some of these practices and the operating philosophy thereof may be applied to grazing of food animals. Unfortunately, despite the magnitude of feed consumed in the latter case, and its importance to high quality protein production in the developing world, what amounts to good manufacturing practices remains to be defined. Available information suggests that the health consequences to man from disease transmitted up the food chain are minimal for manufactured feeds, but may be more significant in subsistence livestock farming including aquaculture. The nature and magnitude of human health problems associated with grazing, particularly in aquaculture, is still being defined for the developing countries of the world. In most cases basic hygiene and plant, or pond sanitation can obviate human health problems associated with animal feeding.

INTRODUCTION

The task of the Consultation is to identify and advise on the control of potential health hazards from foods of animal origin which may originate in formula feeds and grazing. Recent public concerns prompted by the outbreak of bovine spongiform encephalopathy (BSE) in the United Kingdom (UK), and other more common food problems associated with *Salmonella, E. coli* and other micro-organisms, have encouraged health professionals and the feed industry to scrutinise more closely the causes and control of these diseases. Some of the corrective measures are as basic as improving housekeeping and staff training in feed mills. Other means require more difficult challenges of possibly limiting the use of, or radically changing the way some ingredients are prepared (processed), sourced, or where animals are grazed. World-wide the tonnage of feed exceeds 4 billion tonnes per annum of which some 550 million tonnes are milled feeds. The largest portion of the 4 billion tonnes of feed involves subsistence farming on the Indian sub-continent and Asia. Present knowledge of human health as it relates to this sector is at best limited, hence this paper tends to emphasize what is known of the feed industry from the view of the developed world. A large effort needs to be made to define the nature of the impact of aquaculture and other subsistence livestock operations on human health. This activity may represent the world's largest recycling enterprise, employing tens of millions of people. It is an enormously complex global materials handling and manufacturing effort involving the movement of huge quantities of by-products and co-products throughout the world and the extensive movement of animals. Yet despite the magnitude of livestock production, the frequency of human health problems associated with this enterprise is very low. This paper will attempt to identify problem areas and to set forth scientifically reliable procedures to minimise the transmission of hazards from foods of animal origin to human health.

DEFINITIONS

Additive:
An ingredient or combination of ingredients, other than a premix, added to the basic feed mix or parts thereof to fulfil a specific need. Usually used in micro-quantities and requires careful handling and mixing. (1).

Complete feed:
A nutritionally adequate feed for animals other than man which, by specific formula, is compounded to be fed as the sole ration and is capable of maintaining life and/or promoting production without any additional substances being consumed except water. (1)

Compound feed:
A mixture of products of vegetable or animal origin in their natural state, fresh or preserved, or products derived from the industrial processing thereof, or organic or inorganic substances, whether or not containing additives, for oral feeding in the form of a complete feed (see also formula feed). (2)

Concentrate:
A feed used with another to improve the nutritive balance of the total and intended to be further diluted and mixed to produce a supplement or a complete feed. (1)

Feed (feedingstuff):
Any substance, whether processed, semi-processed or raw which is intended for animal consumption. (1,3-5)

Food:
Any substance, whether processed, semi-processed or raw which is intended for human consumption, including drinks, chewing gum and any substance which has been used in the manufacture, preparation or treatment of 'food' but excluding cosmetics, tobacco and substances used only as drugs.

Formula feed:
Two or more ingredients proportioned, mixed and processed according to specifications (see also compound feed). (1)

Hazard:
A biological, chemical or physical agent in, or a property of, feed which may have an adverse effect (6).

Ingredient:
A component part or constituent of any combination or mixture making up a (commercial) feed (1,4)

Medicated feed:
Any feed which contains drug ingredients intended for the treatment or prevention of disease of animals other than man. (Note: Antibiotics used as growth promoters are usually considered to be 'feed additives') (1)

Premix:
A uniform mix of one or more micro-ingredients with a diluent and/or carrier. Premixes are used to facilitate uniform dispersion of micro-ingredients in a larger mix or a mixture of additives, or a mixture of one or more additives with substances used as carriers, intended for the manufacture of feed (1).

Risk:	A function of the probability of an adverse effect and the magnitude of that effect, consequential to a hazard(s). (6)
Risk analysis:	A process consisting of three components: risk assessment, risk management and risk communication. (6)
Risk assessment:	The scientific evaluation of known or potential adverse health effects resulting from exposure to a hazard. The process comprises the following steps: (i) hazard identification, (ii) hazard characterisation, (iii) exposure assessment and (iv) risk characterisation. The definition includes quantitative risk assessment and qualitative expressions of risk, as well as an indication of uncertainties. [adapted from (6)]
Risk management:	The process of weighing policy alternatives to accept, minimise or reduce assessed risks and to select and implement appropriate options. [adapted from (6)]
Straight feedingstuff or straights:	A vegetable or animal product in its natural state, fresh or preserved, and any product derived from the industrial processing thereof, and single organic or inorganic substance, whether or not it contains any additive, intended as such for oral feeding (2).
Supplement :	A feed used with another to improve the nutritive balance or performance of the total and intended to be (1):

(a) Fed undiluted as a supplement to other feeds;

(b) Offered free choice with other parts of the ration separately available, or;

(c) Further diluted and mixed to produce a complete feed.

CONTROL OF FEED PREPARATION, MANUFACTURE AND DISTRIBUTION

Introduction

The quality of livestock feed and forage and their potential impact on human health begins with the growing and harvest of feedstuffs in the farmer's field and/or the grazing of the animals. It has already been mentioned that the size of this extremely diverse enterprise is some four billion tonnes turnover per annum, of which the majority is on subsistence farms on the Indian subcontinent and Asia. Feedstuff quality is affected all along the, sometimes, lengthy market route to the consumer of animal products. It is wise for the feedstuff (commodity) user to know that the ingredients being purchased for feed, or the area being grazed, is free from contamination which would not ordinarily be removed by processing, and/or that pastures and ponds are free from pollution or other contamination. Be it the large-scale full-line feed mill producing finished feeds for sale, or the small on-farm feed mixer, the quality of the ingredients is of importance to the health of the animal consuming the feed and to the human consumer who uses the animal products. The buyer of these raw materials should know that the feedstuffs being bought have come from sources where the feedstuff is handled in such manner as to minimise exposure to moisture, pests, toxic chemicals, filth, microbial or other contamination which could cause health problems in food animals and subsequently in human consumers. Training of workers at all levels of the handling and processing by which feedstuffs become animal feed is important to the maintenance of a healthy feed supply. The vehicles, vessels, storage facilities, conveying equipment and environmental management should all be maintained at the highest standard of cleanliness and free of excess moisture so that spoilage is controlled and the conditions under which contaminants such as mycotoxins and *Salmonella* flourish are effectively eliminated. On-going sampling of ingredients to be certain that quality standards are met and testing for any suspected contaminants, plus a constant effort at good housekeeping, will minimise health problems attributable to the feeding of livestock. Regarding grazing and natural forages, such as free-ranging cattle, as well as enrichment of aquaculture ponds with animal manure, measures should be taken to assure that the forages and ponds are not contaminated with heavy metals, pesticides, radionuclides (in those regions where this is a known problem), or mycotoxins. For the aquatic food animals sanitary measures should be taken to avoid infecting workers or consumers of these products with the bacterial, viral and parasitic agents found in night soil and manure. The means exist to minimise the contamination of the manufactured feeds through appropriate technology, that is, through current good manufacturing practices (CGMP), and careful handling of the feedstuffs. Equally, the application of good aquaculture management and sanitation practices in the ponds and raceways should reduce the possibility of infection to workers and consumers of aquaculture products, particularly for freshwater species. In the latter case, good aquaculture management, plus treatment of manure and other by-products, and increasing use of finished feeds should, over time, reduce the likelihood of human health problems.

Feed and Forage Quality Assurance (7)

Quality assurance (QA) begins with the concept of what the feed product is to be, in terms of the species being fed and the results being sought. Ingredient specifications are important to quality assurance in defining the quality of the feedstuffs to be accepted by the processor when the raw materials are received for processing. The formulation of the finished feed, including any added medications, should meet the regulatory requirements of the government as well as satisfy the animal production objectives of the customer. Other QA factors involve the manufacture and distribution of the feed. Not only should the feed be of good quality, it should also arrive in good condition and in a timely fashion. The key elements in effective quality assurance (8) at the feed production facility should include the following.

Proper sampling

Proper sampling of ingredients at receipt should be carried out as appropriate and in accordance with AOAC International procedures (9).

Laboratory testing and microscopy

Laboratory examination may be indicated if contamination is suspected, or for routine samples (4,8,10) to determine compliance with contract specifications. (Ingredients or feed samples can be taken in remote areas and mailed for examination at many qualified laboratories throughout the world. Often the results can be returned via fax in a few days. Shipment of such samples needs to be made within the bounds of local and destination quarantine regulations).

In-plant quality control

In-plant quality control includes monitoring product as it is produced to be certain it meets formula specifications. Such may include visual and laboratory tests (3).

Control of drug carry-over

Clean-out of equipment between batches to prevent cross contamination (to avoid contamination of another feed) is an important QA procedure and typically a government requirement (8).

Plant sanitation and integrated pest management

Sanitation and control of pests are an extension of quality assurance wherein the feed manufacturer controls the entry of potential health hazards into the manufacturing process. From the point at which feedstuffs are received (the receiving pit, elevator, etc.) it is the processor's responsibility to minimise the presence of substandard or contaminated ingredients in the plant. This means inspecting and sampling incoming loads when contamination (mouldy or wet, or insect-ridden) is suspected. High moisture levels can encourage the growth of bacteria, fungi and moulds, thereby resulting in the presence of potentially harmful disease organisms and mycotoxins (1,3,5,11). Likewise the fouling of the feedstuffs and finished goods by birds and rodents can be the vector by which the feed is potentially contaminated by pathogens such as *Salmonella*. The integrated approach requires that there be regular inspections of the plant and the inbound feedstuffs; that there be good housekeeping; that physical and mechanical methods are applied to keep pests out of the plant; and that chemical applications (fumigation for insects or propionic acid/formic acid treatments for bacteria and fungi) are used correctly.

Plant cleanliness

The cleanliness of a feed manufacturing facility should start with the design of the plant. All surfaces should be accessible; there should be no dead space. The plant and its equipment should be accessible and easy to clean and the facility perimeter relatively free of debris and undergrowth. The exterior of the plant should be reasonably clean and free of filth, including dust. Cleaning is a routine activity, hourly or daily, to prevent the accumulation and spoilage of spilled feeds, feedstuffs, or other components used in feed manufacture.

The receiving area

The location at which raw materials are received should have adequate dust control and also be easy to clean to prevent attraction of pests, particularly birds and rodents which may carry *Salmonella* or other pathogens which would be carried with the feed to the animal and possibly the ultimate consumer. In tropical and some temperate climes, contamination by fungi, moulds and bacteria may be controlled, in addition to good housekeeping, by such means as running a couple of tonnes of barley containing 5% propionic acid through the conveying systems of the mill. (The barley can be saved for reuse several times.) Care should be taken with propionic acid or formic acid due to the corrosive effects on the machinery and palatability problems - there are other alternatives. With careful application these acids or their salts (less corrosive) can reduce microbial counts (1).

24

Storage

Tanks, augers and conveyors should be designed so they are easy to clean and to minimise accumulation of spoiled ingredients or other contaminants, including drugs or ingredients from previous batches or loads. Housekeeping ought to be a regular part of the operation of all plant facilities. Tanks containing ingredients should be monitored for temperature and moisture, especially in the elevated temperatures and humidity of the tropics. Elevated temperature and moisture levels are an early sign of potential deterioration in feedstuffs and/or finished feeds, due to fungal or insect infestations, or both.

Prevention is the best measure

Processing will not remove mycotoxins, heavy metals, and some pesticides, but processing which includes pelleting, extrusion, or otherwise heat-treating the feed can kill or significantly reduce the number of bacterial pathogens. Further along the process stream, the handling systems can re-contaminate such feed before it reaches the animal to be fed, if these areas are not routinely cleaned as noted above. By denying access by pests or other contaminants to the plant (rodents birds, pesticides, etc.), the cleaned feedstuffs, ground mash, pelleted or extruded feed, which are relatively clean, are less likely to be reinfested.

Handling of finished feeds: their storage and transport

Bulk storage

Storage bins ought to be cleaned routinely to prevent cross-contamination of ingredients and the accumulation of spoiled feed, the latter of which, as mentioned above, can infest otherwise clean feed with insects, fungi or filth.

Warehousing

Buildings where feed is stored should be clean and secure from pests including rodents and birds. Sacked feeds should be stacked with a gap between them on pallets and off the floor to allow adequate cleaning, ventilation and inspection.

Transportation

All transport vehicles should be free of contaminants so that cross-contamination, or contamination with pests or other cargoes is minimised. Plant management should inspect all contract vehicle carriers, as well as its own fleet, to be confident that the means of transport does not create a health problem for the animals or the human consumer. Contamination of fresh feed from fertilisers, chemicals, moisture or from other previous cargoes is a potential concern, as is the possibility that recycled bags which have toxic residues may be used.

Current Good Manufacturing Practices

CGMP includes the material discussed above in the total manufacturing context (3,8). Record keeping should be an integral part of the receiving and processing functions. This permits claims to be made against suppliers for defective ingredients and provides information so that any defective feed that has been sold can be recalled, or the consumer warned of the defects in manufacture. The fact that much of the world follows CGMP at some level accounts for the fact that diseases affecting human consumers of animal products are rare. The checks and balances of the CGMP system allow tracking as well as analysis and action to prevent problems before they affect the human consumer of livestock products. Such management and manufacturing controls can be developed for subsistence farmers and others in remote areas through an extension agent system.

Good Grazing Practices

Grazing livestock including aquatic species should not be put at risk on lands or water exposed to agricultural spray drift or other industrial or naturally toxic events or activities which could introduce toxins or diseases. Field and laboratory testing of forages and water analysis plus a thorough knowledge of the land to be grazed should be a routine part of animal production in any geographical location.

REFERENCES

1. McEllhiney, R.R. Technical Editor. 1994. *Feed Manufacturing Technology IV.* Arlington, Virginia. American Feed Industry Association. 606 pp.

2. HMSO. 1992. The Report of the Expert Group on Animal Feedingstuffs to the Minister of Agriculture, Fisheries and Food, the Secretary of State for Health and the Secretaries of State for Wales, Scotland and Northern Ireland. London. Her Majesty's Stationery Office.

3. FAO/WHO. 1996. Codex Alimentarius Commission: *Proposed Draft of Hygienic Practice for the Products of Aquaculture.* FAO, Rome.

4. Association of American Feed Control Officials (AAFCO). 1996. The AAFCO Official Publication: 108-110. Atlanta, Georgia. AAFCO.

5. Meronuck, R. & Concibido, V. 1996. *Feedstuffs Reference Issue: Mycotoxins in Feed*: 139-145. Minnetonka, Minnesota. Miller Publishing.

6. FAO. 1995. Expert Consultation Report: *Application of Risk Analysis to Food Standards Issues.* Rome

7. Council for Agricultural Science and Technology. 1994. Foodborne Pathogens: Risks and Consequences. *Task Force Report No. 122.* Ames, Iowa.

8. American Feed Industry Association. 1993. *Model Feed Quality Assurance Manual*: 1-26. Arlington, Virginia.

9. AOAC International. 1995. *Official Methods of Analysis*, 16th Ed.: 4.1.01. Gaithersburg, Maryland.

10. FAO. 1987. Proceedings of the FAO Expert Consultation on Substitution of Imported Concentrate Feeds in Animal Production. *Animal Production and Health Paper No. 63.* Rome

11. Blackman, J., Bowman, T., Chambers, J., Kisilenko, J., Parr, J., St. Laurent, A-M., & Thompson, J. 1990. *Controlling Salmonella in Livestock and Poultry Feed* Agriculture Canada: 1-20. Ottawa.

INFECTIONS AND INTOXICATIONS OF FARM LIVESTOCK ASSOCIATED WITH FEED AND FORAGE

SUMMARY

For the purpose of this paper, animal feed includes any substance, whether processed, semi-processed or raw which is used for animal consumption. It includes, therefore, forage crops, manufactured feed and such things as animal and human wastes. Forage comprises green plants, including the ear or seed head, which may be consumed by animals, either fresh or as cured or fermented product. The term food is confined to any substance, whether processed, semi-processed or raw which is consumed by humans.

Animal feed or forage may be the source of a limited number of infections for farm animals that could in theory lead to human illness. These include *Salmonella enterica* and *Toxoplasma gondii, Trichinella spiralis* and possibly the agent of bovine spongiform encephalopathy (BSE). The risk to human health from several other infectious agents, which may contaminate either feed or forage, appear to be either negligible or non-existent. These include *Bacillus anthracis*, *Clostridium botulinum* toxin, *Listeria monocytogenes* and *Mycobacterium bovis*.

Animal and human waste may be incorporated in animal feed or can be used to fertilise forage crops. The use of untreated human wastes in fish farming may be associated with serious human health problems. For example, liver fluke infestation (clonorchiasis and opisthorchiasis) in Southeast Asia.

Mycotoxins in animal feed can result in foods of animal origin containing these compounds. This risk is well recognised but it has yet to be quantified accurately and in some instances the risk may be of theoretical rather than practical importance.

Pesticides, agricultural and industrial chemicals, heavy metals and radionuclides may pollute animal feed and forages. The methods available for controlling pollution from these sources are well understood from a technical viewpoint although the effective implementation of controls can prove to be difficult.

INTRODUCTION

This paper considers the causes of adverse effects that can occur in humans as a consequence of contamination of food through the feeding of animals with contaminated feedingstuffs. The animals of concern are (a) farmed livestock including farmed game, particularly ruminant species (cattle, goats and sheep), pigs and poultry and (b) farmed fish, particularly fresh water species kept in small water bodies or ponds.

Each year many million tonnes of forages and manufactured animal feed are consumed by animals worldwide but fortunately this results in few serious human infections, diseases and intoxications. Pathogenic agents involved may be infectious or non-infectious and can be divided arbitrarily into five groups (see Table 1). Of these, three categories (Nos 2, 3 and 4), are of no direct relevance to this Consultation and will not be considered in this document.

Table 1. Pathogenic agents of animals associated with feeding

	Feed or feed ingredients	Dried or fermented forages	Pasture or or grazing land	Waste food or garbage
1. Infectious agents transmissible to humans from farm animals i.e. zoonoses	*Bacillus anthracis* spores Bovine spongiform encephalopathy agent *Salmonella enterica* Newcastle disease virus*	*Toxoplasma gondii*	*Bacillus anthracis* spores *Mycobacterium* spp. From wildlife sources Cestode eggs e.g. *Cysticercus bovis*	
2. Non-zoonotic infectious agents, or their products, which cause disease in both farm animals and humans		*Clostridium botulinum* toxin *Listeria monocytogenes*		
3. Infectious agents causing epidemic diseases in farm animals which may result in human hardship	Viruses of African swine fever, foot and mouth disease and swine fever.			Viruses of African swine fever, foot and mouth disease and swine fever.
4. 0Non-infectious agents which cause disease in farm animals and humans	Fungal hyphae and spores causing allergic disease	Fungal hyphae and spores causing allergic disease		
5. Products of non-infectious agents which cause disease in farm animals and humans	Mycotoxins	Mycotoxins	Mycotoxins	

29

* The virus of Newcastle disease may be spread to poultry via animal feed and from birds to humans via aerosols. The risk of infection is very low. In humans the virus may cause, for a few days, very mild conjunctivitis and symptoms similar to a head cold. There is no risk to humans from poultry meat.

In recent years public concern on health matters associated with food has increased as a consequence of, *inter alia*, the outbreak of bovine spongiform encephalopathy (BSE) in the United Kingdom. Also of great concern are the world-wide epidemic of *Salmonella* serotype Enteritidis food poisoning (poultry, meat and eggs) and the more localised outbreaks of illness associated with *Listeria monocytogenes* (dairy products, paté and salad crops) and *Escherichia coli* O157:H7 (ground or minced beef). The appearance of problems such as BSE and Enteritidis food poisoning, has resulted in the enacting of specific legislation in several countries, while the general heightening of interest world-wide has prompted health professionals and the feed industry in many countries to scrutinise the control of potential pathogens and to formulate better procedures for purchasing and handling of feed ingredients and the subsequent manufacture of compounded or formula feeds.

This review has been prepared following a detailed evaluation of the scientific literature. The principal database consulted was the Science Citation Index 1981-1996. It was intended to provide, when appropriate, a balanced assessment of the situation in both developed and developing countries. However, this was frequently not possible as the published information concerning the latter was either very slight or non-existent.

DEFINITIONS

For the purposes of this paper the following definitions have been used (in addition to those noted in Annex 3). The source of some of them is acknowledged although some have been modified slightly to make them relevant to this document.

Endemic:	Habitually present in a population. Often used to describe certain diseases (1).
Epidemiology:	The study of disease, and disease attributes, in defined populations (1).
Farmed livestock:	Ruminant species (principally cattle, goats and sheep), pigs and poultry
Forage:	Green plants, including the ear or seed head, which may be consumed by animals, either fresh or as a cured (e.g. hay) or fermented (e.g. silage) product.
Zoonosis:	A disease common to man and other animals, usually in which animals are the main reservoir of infection (1).

HEALTH HAZARDS ASSOCIATED WITH ANIMAL FEED AND FORAGES

A joint FAO/WHO Expert Consultation on the 'Application of Risk Analysis to Food Standards Issues' was held in Geneva during 1995 and the report of this meeting contains much of relevance to animal feed (2). Among other things the consultation recommended that risk assessment and risk management should be considered separately and it is the former topic which is of principal concern in this paper.

That consultation concluded that the estimation of risk from chemical and physical hazards in feed is an incompletely developed science although it is generally more advanced than that for biological agents. This represents a serious deficiency since the consultation considered that biological risks are probably greater, and present more immediate problems to human health. Attempts to quantify risks are hampered by a considerable degree of uncertainty and it is essential that this is taken into account when protocols for risk management are being formulated, particularly because these may have considerable financial implications for the agricultural and feed industry.

Diseases and intoxications in farm livestock associated with the consumption of feed and forages have been reviewed (3-5) while more recently there have been reviews on specific problems including BSE (6-8), the *salmonellae* (9) and mycotoxicosis in poultry (10).

In general it can be assumed that any problem involving feed will pose a greater risk to the animal that consumes it than the consumers of meat, milk and eggs. The risk may take two principal forms, either vegetative bacterial cells or other microorganisms which colonise the animal following consumption of the feed and which persist to contaminate, and possibly multiply on, the food product (usually meat or milk and more rarely eggs), or toxic microbial metabolites which are either excreted in the milk or eggs or which persist as a residue in the meat. The classification of prions associated with BSE and scrapie remains uncertain and they may subsequently form a separate category.

Unlike most chemical and physical hazards, the vegetative microbes pose a special problem in assessing risk since their numbers may increase if environmental conditions are favourable, thereby increasing the risk to the consumer during storage, handling and further processing. Risk assessment depends, *inter alia*, on being able to quantify the numbers of microorganisms present or to measure the concentration of toxic metabolites such as mycotoxins. Many methods are in use for the enumeration of microorganisms in laboratories around the world but the lack of standardisation means that it is difficult to compare directly the results obtained by different laboratories. International Organization for Standardization (ISO) methods exist for several of the principal pathogens e.g. the *salmonellae* although these methods are frequently complicated and may be too expensive for the routine laboratory. There has been great interest in the development of rapid methods. Several different approaches have been proposed for assessing bacterial numbers but they may prove less sensitive than 'traditional' microbiological methods and more expensive. These include direct epifluorescent microscopy and electrical methods such as impedance and conductance measurements of fluid bacterial cultures. Methods based on DNA technology may be useful for identifying microorganisms and for studying their epidemiology but are generally unsuitable for their enumeration.

Infectious agents and their products associated with feed and forage

The principal biological agents presenting a risk to the public health, and which may be transmitted by animal feed, have been summarised in Table 1. These include:

Bacillus anthracis

Anthrax, which is caused by *Bacillus anthracis*, has a world wide distribution although it occurs only sporadically in temperate countries. *B. anthracis* sporulates on exposure to air and the resulting spores are capable of surviving for long periods in the environment and in contaminated animal feed, particularly meat and bone meals prepared from animals which have died from anthrax. For a review on spores see Dragon and Rennie (11). Anthrax is principally a disease of cattle and is only rarely identified in other farm livestock. In cattle it is associated with sudden death although it is possible to diagnose the illness in the live animal and this can be treated with penicillin. Vaccines are available for human use in areas where the disease is endemic, however it is unlikely that affected animals represent a direct risk to humans. This assertion is supported by the fact that there is little evidence from published medical statistics that humans contract anthrax from meat and milk, although there have been cases of anthrax in people who have consumed meat from animals which have died of anthrax. Conventional heat treatment may not eliminate spores, hence, bone meals prepared for use as fertilisers have been implicated as a source of anthrax spores for people handling these products. However, an understanding of the ecology of the bacterium, coupled with improvements in industrial hygiene practices in many countries, has resulted in a considerable reduction in the incidence of anthrax as an occupational disease of people working with animal products such as hides, wool, meat and bone meal.

Transmissible spongiform encephalopathies

Bovine spongiform encephalopathy (BSE)

BSE, a non-febrile neurological disease of bovine animals with a long incubation period, was first diagnosed in the UK during 1986. The disease was made notifiable in the UK during 1988 and the legislation was updated with a new Bovine Spongiform Order during 1996. A WHO Consultation on the public health issues related to human and animal transmissible spongiform encephalopathies (TSE) was held in Geneva during April 1996 (12). A summary of its recommendations for the protection of public health follows:

- No part or product of any animal which has shown signs of a TSE should enter any food chain (human or animal). In particular all countries must ensure the killing and safe disposal of all parts or products of such animals so that TSE infectivity can not enter the food chain and should review their rendering procedures to ensure that they effectively inactivate TSE agents. It must be recognised that the BSE agent is remarkably resistant to physico-chemical procedures which destroy the infectivity of common microorganisms.

- All countries should establish continuous surveillance and compulsory notification for BSE. In the absence of surveillance data the BSE status of a country must be considered unknown.

- All countries should ban the use of ruminant tissues in ruminant feed.

- Milk and milk products are considered safe. Gelatin and tallow are safe if manufactured in a manner which involves the inactivation of residual infectivity.

- More studies are required to allow a full risk assessment. Incomplete risk assessment hinders accurate risk communication and perception.

- Bovine materials destined for the pharmaceutical industry should be obtained from countries which have a surveillance system in place and which report either no or only sporadic cases of BSE.

- Research on TSE should be promoted, especially regarding rapid diagnosis, agent characterisation and epidemiology.

Detailed epidemiological and other studies, which have been summarised by various authors (6-8,13), identified a link with the consumption by young calves, of feed contaminated with a scrapie-like agent, derived from sheep or cattle. The BSE agent, or prion, is very resistant to heat and will survive the temperatures normally used for processing animal tissues prior to their inclusion in animal feed. It has been suggested that changes in the processing of animal tissues during the late 1970s primarily introduced to increase protein yields due to less denaturisation, and to conserve energy may have precipitated the epidemic. The rapid growth of the epidemic from 1986 to 1991 is consistent with recycling of contaminated material in animal feed (8). A scheme for eradicating the BSE agent was introduced in the UK during 1996. This involved the slaughter of all cattle over 30 months of age and the disposal of their carcasses by burning or rendering. This drastic policy has been questioned by several authors (8) who undertook a detailed and comprehensive statistical assessment of the available epidemiological data. These authors concluded that the epidemic is well past its peak, that new infections from animal feed should have ceased to occur by the end of 1994 and that the epidemic will die out naturally by the year 2001 because the small numbers of infections contracted by calves *in utero* will be insufficient in themselves to maintain the epidemic.

The British Government introduced measures to control the feeding of bovine offals to cattle during 1988 and these have since been modified. The present legislation entitled 'The Specified Bovine Material (No. 3) Order 1996' came into force in July 1996 and imposes stringent control measures on the slaughter industry. The Order is implemented to a large extent by the Meat Hygiene Service, an executive agency of the Ministry of Agriculture Fisheries and Food.

The objective of the Specified Bovine Material (SBM) Order was to prevent various bovine tissues, and other waste which may be infected with the BSE agent, from entering the food chain of farm animals and man. The complexity of the provisions indicate that this apparently simple objective is no easy task. The staining of SBM facilitates its identification after it leaves the abattoir. The Order makes a distinction between animals over 6 months of age and those which are slaughtered when over 30 months, under the slaughter scheme introduced in compliance with Commission Regulation (EC) 716/96 (the so-called "scheme animals"). An ELISA test has been developed in the UK to detect mammalian protein in animal feed and is now in regular use in the BSE control programme.

The Order requires that (a) SBM must not be sold for consumption by animals including humans, (b) premises producing mechanically recovered meat must be registered, (c) the vertebral column (including the sacrum but not the coccygeal vertebrae) must not be used in the production of mechanically recovered meat, (d) the head, after removal from the carcass and after removal of the tongue, and the other SBM, must be sprayed or immersed in Patent Blue V (0.5% w/v) so that all surfaces are covered and then kept separately from all other animal material, (e) the brain and eyes shall not be removed from the carcass of a bovine animal aged over 6 months while the spinal cord can only be removed in a slaughterhouse prior to disposal in an approved manner and (f) if whole carcasses are rendered they must be treated as SBM.

SBM, which must be transported in impervious covered containers, can only be removed from a slaughterhouse to an approved collection centre, incinerator or rendering plant; a diagnostic or research laboratory; or an approved premises not concerned with the manufacture or preparation of food or feedingstuffs.

Schedule 1 of the Order provides the requirements for rendering plants. The provisions of Part 1 require the separation of untreated and treated SBM, the provision of equipment with continuous recording devices to measure temperature and where necessary pressure and a safety system to prevent inadequate heating.

The question as to whether BSE is a zoonosis remains a matter of debate although prudence dictates that a cautious attitude is taken over the issue. Recent studies (14) provide evidence of a similarity between the BSE agent and that which causes new variant Creutzfeldt-Jakob disease (NV-CJD) in humans although further work is required to prove this conclusively. The problem is further compounded by the fact that there is little uniformity in the definition and reporting of suspect cases of Creutzfeldt-Jakob disease (CJD) in different countries. This has been highlighted in a recent survey in the European Union (EU) (15).

The occupational risks from the BSE agent for people at work has been considered by the UK's Advisory Committee on Dangerous Pathogens and they concluded, *inter alia*, that there is no evidence of any risk to those in occupations in which exposure to the BSE agent may occur. However, as BSE is apparently a new phenomenon, it is prudent to take precautions where there is risk of exposure.

Scrapie

Another important issue concerns the scrapie agent which affects sheep and to a lesser extent goats. Briefly, scrapie is a natural infection of adult sheep which is transmitted from the ewe to the lamb either before or soon after parturition. The agent can also spread horizontally between unrelated sheep. Scrapie can be controlled by selective culling in the female line and by the introduction of husbandry measures which limit horizontal spread of infection at lambing. This approach is a long-term solution however, since it is necessary to build-up accurate records over several years in order to identify the ewes which need to be culled (13).

In 1996, the EU Scientific Veterinary Committee recommended that offal from sheep and goats should be banned from the farm animal and human food chain. This proposal was rejected by the Standing Veterinary Committee and subsequently by EU farm ministers during a meeting held during December 1996.

Transmissible mink encephalopathy (TME)

TME is a very rare disease of ranch-reared mink. The disease is thought to be caused by an exogenous source of infection, possibly via contaminated feed containing abattoir waste derived from sheep or goats. TME is a 'dead-end' disease with no natural routes of transmission from mink to mink (13).

Feline spongiform encephalopathy (FSE)

FSE was first reported in the UK by veterinarians working at Bristol University (16,17). A further case has been reported in January 1997, in a cat which was born in the UK after the extension of the ban of SBO during 1990. This, according to a government spokesman, provides further evidence that BSE could cross the species barrier. The most likely source is commercial cat food rather than raw tissue or rendered products.

Salmonella enterica

There are over 2000 *salmonella* serotypes and these can be divided arbitrarily into three unequally sized groups. These include, the species specific serotypes such as Dublin (cattle) and Gallinarum and Pullorum (poultry); the 'invasive' serotypes which may cause septicaemic disease in several animal species (e.g. Enteritidis and Typhimurium); and the 'non-invasive' serotypes. The third group is by far the largest and may be associated with subclinical infections in farm livestock. They can cause disease on occasions and are associated with food poisoning in humans.

Animal feed ingredients, of both animal and plant origin, are frequently contaminated with *salmonellae* although the most common serotypes isolated are rarely the most prevalent in animals including man. The two most important serotypes associated with human disease, Enteritidis and Typhimurium, are rarely isolated from animal feed. Several different methods have been recommended for the isolation of *salmonellae* from feed. The efficiency of any cultural method may be influenced by the presence of antibacterial agents in the feed (18).

Salmonellae are widely distributed in nature and animal feed is only one of many sources of salmonellas for farm animals. The control of *salmonella* infections therefore requires a comprehensive approach in which feed control is one component of a overall control programme. The risk factors in relation to *salmonella* infections in pigs have been investigated (19).

The most effective means of eliminating salmonellas from feed is pasteurisation (e.g. 85°C with a residency time of four minutes) although it may not always be appropriate to do this for all feeds. For high risk products such poultry feed, compulsory heat treatment may be justified, although care must be taken to prevent subsequent recontamination. Irradiation will also effectively kill salmonellas in feed (20) and may prove a practical proposition for elite breeding flocks. The expense will preclude its use for commercial feed.

Growth promoters and salmonella shedding : Over the years there has been much debate about whether the inclusion of growth promoting antibiotics (agents which are generally active against Gram positive bacterial genera) in feed favour the colonisation of the intestinal tract with salmonellas. There has been relatively little published on this topic in recent years, particularly in respect to salmonella carriage in birds reared on commercial farms. However, the problems caused by a lack of veterinary supervision of the use of growth promoting antibiotics has been highlighted recently (21). One author (22) reviewed the results of experimental studies and concluded that shedding was prolonged in some but not others. The majority of investigations referred involved the study of broiler birds that were inoculated orally or were reared in contact with orally inoculated birds. There is relatively little information, however, concerning challenge *via* feed, which is an important potential source of salmonella organisms for birds reared under commercial conditions. It has been reported (23) that salmonella shedding was increased in birds given feed containing avoparcin and contaminated by the inclusion of unsterilised meat and bone meal while the results obtained using a naturally contaminated feed, were inconclusive (24). Further studies involving young chicks given feed artificially contaminated with salmonellas and containing either avoparcin, virginiamycin and zinc bacitracin have since been reported (25). The principal conclusion drawn from these was that the presence of the growth promoter was associated with an increase in the prevalence of salmonella colonisation of between 15-20% and that the average number of salmonellas in faecal contents of the positive birds was increased by <10-fold.

Overall, it was calculated that the total numbers of salmonellas within a group of birds was within the same order of magnitude. This means that, in microbiological terms, the growth promoter did not lead to a 'super' infection in the intestinal tract.

Mycobacterium spp.

Wildlife species e.g. the badger (*Meles meles*) in the UK and brush-tailed possum (*Trichosurus valpecula*) in New Zealand are both reservoirs of infection with *Mycobacterium bovis*. In theory these animals can pollute pasture land with faeces and urine. Cattle contracting tuberculosis from these sources pose a theoretical risk to the human population although this is unlikely since cattle in areas where the problem is recognised, e.g. southwest England, are tested on a regular basis and reactors are slaughtered at an early stage in the infection and before they develop tuberculous pneumonia or mastitis.

Tuberculosis remains a problem in cattle in many developing countries. The principal problem to be addressed is to eradicate the infection from the cattle and hence reduce the risk of infection spreading to the human population. When eradication reaches an advanced stage, epidemiological studies may reveal that persistent infections in 'problem herds' may be associated with wild life reservoirs which, like the badger, pollute the pasture land. Clearly these problems will need to be addressed as and when they are identified.

Toxoplasma gondii

The protozoan *Toxoplasma gondii* may cause abortion in pregnant ewes and neonatal deaths in lambs. A potential source of infection in ewes is preserved forage or bedding contaminated with the faeces of cats and rodents, the principal hosts of this coccidian parasite (26). The elimination of cats and rodents from the farm environment is an obvious control measure although probably unachievable in practice.

The prevalence of seroconversion in the human population can be very high (27). Humans may become infected by accidental ingestion of oocysts from cat faeces due to poor hygiene practices. In addition, the cysts of the parasite may be present in the muscles of meat animals, particularly sheep (28), and they pose a threat to human health if the infested meat is eaten raw or partially cooked.

Trichinella spiralis

Trichinella spiralis is principally a parasite of pigs. Viable cysts in the muscles act as a source of infection for humans who consume raw or partially cooked meat. The cysts are temperature sensitive and can be killed by freezing and cooking temperatures.

Control of animal pathogens in feed

<u>Heat treatment</u>. The pasteurisation of feed ingredients by heating to 80-85°C for four minutes will eliminate most vegetative bacterial cells and these temperatures can be readily applied during the manufacture of compound animal feeds. Bacterial spores and prions e.g. those of *B. anthracis* and the BSE agent are not destroyed by these temperatures, however. To control prions it is necessary, within the EU, to heat mammalian animal wastes to >133°C for 20 minutes at a pressure (absolute) of 3 bar with a maximum particle size of 50 mm.

Fermentation. The composting of animal wastes prior to their inclusion in animal feed will reduce the numbers of vegetative bacteria but will not eliminate spores. The process can be rendered more effective by the addition of fermentable carbohydrates e.g. molasses and/or organic acids e.g. propionic acid. The development of an anaerobic digester has been described (29), which converts animal waste to methane and destroys pathogens.

Infectious agents associated with farm animal and human waste used as manure

Farm animal and human excrement, either raw or treated, may be spread as a fertiliser onto land or into water. This procedure presents a number of risks to the health of farm livestock and fish and some of the animal disease agents may be transmissible to man.

Animal and human faecal waste is a valuable resource for the developing world. Ideally it should be properly treated before being applied to land or water. However, such treatment may not be available and the benefits obtained in enhancing food production may outweigh the risks to health. This problem was addressed by the Codex Committee on Fish and Fishery Products in 1996. The Committee recommended that fertilisers of natural origin should only be used if they do not compromise the hygienic quality of fish from an epidemiological, chemical, microbiological or parasitological standpoint. These fertilizers may comprise animal manure, stable drainage, slaughterhouse waste, nightsoil, excreta-derived sludge and seepage, and municipal sewage. Raw animal and human wastes are not suitable for pond enrichment since they may contaminate fish with pathogenic organisms.

Dosage, frequency and timing of fertilisation should be adjusted so that there is no deterioration of the pond environment and the health of the fish and their hygienic quality is not affected. It is recommended that the application of properly composted organic fertilisers is stopped for an appropriate period prior to harvesting the fish.

Waste spread to land

Excrement and manure are useful fertilisers, particularly in developing countries with impoverished soils (30,31). However, these materials are a source of large amounts of ammonia and methane which cause atmospheric pollution (32-37) while nitrates and other inorganic compounds may pollute ground water (38). In addition, there are a number of potential risks to the health of animals including man following the disposal of sewage sludge onto farm land. The harmful components include microorganisms and heavy metals. This topic has been reviewed (39-41).

The majority of pathogenic microorganisms present in animal wastes pose no risk to human health, with the principal exception being the salmonellas. It has been suggested that, *Mycobacterium paratuberculosis*, the cause of Johne's disease in cattle, may be associated with Crohn's disease in humans. This matter remains one of debate although at present there is no definite proof that *M. paratuberculosis* is a zoonosis (42,43).

The appropriate treatment of manure and sewage can remove most but not all pathogenic microorganisms. Those that do persist may be inhaled in aerosols during spreading or may be ingested following the consumption of salad and other crops which are consumed raw. The health risks to farm animals from grazing pastures treated with slurry have been investigated in great detail. It appears that the risk from infectious agents is very slight since there are few reports of disease problems associated with slurry spreading (44-46) although it is recognised as a significant risk factor in *Mycobacterium paratuberculosis* infection in cattle

(47). That there is some risk is not in doubt, however, and it has been recommended that slurry should be stored for at least a 60 days in summer and 90 days in winter before being applied to pasture. If possible slurry should be spread to land that is not to be used for either grazing or the production of crops eaten raw (e.g. lettuce). If a treated pasture is to be grazed, a further month should elapse before animals are allowed access (44).

Waste dispersed in water

Animal waste may contaminate water supplies and it is possible that the oocysts of *Cryptosporidium parvum* originating from farm livestock may contaminate the water supply. Conventional treatment of water does not eliminate the oocysts. In the event of pollution, water should be boiled before consumption. Conversely the contamination of pasture land with human sewage has resulted in cattle becoming infected with 'human' pathogens such as *Salmonella paratyphi*.

The use of human waste and waste water in aquaculture has been reviewed (48,49). Three potential health risks are associated with the reuse of excreta in ponds. These include (a) passive transfer of pathogens by contaminated fish, (b) transmission of helminths in which fish are the intermediate hosts and (c) transmission of helminths in which other pond fauna are the intermediate hosts (50).

There are several categories of people who are at risk from human waste. These include those that collect and distribute the waste; farm and harvest the fish; and distribute and consume the carcasses. Four main categories of infectious agent are considered to be of relevance to aquaculture (48). These are non-bacterial infections spread by the faecal-oral route; bacterial infections spread by the faecal-oral route; water-based helminths; and insect vectors associated with excreta. These are summarised in Table 2.

Table 2. Recommended maximum levels of heavy metals and halogens in animal feed [adapted from (51)]

Category	Maximum concentration (ppm)	Metal
Highly toxic	10	Cadmium, mercury, selenium
Toxic	40	Barium, cobalt, copper, lead, molybdenum, tungsten, vanadium
Moderately toxic	100	Antimony, arsenic, iodine, nickel
Slightly toxic	1000	Aluminium, boron, bromine, bismuth, chromium, manganese, zinc

Fish are the intermediate host of several parasites of humans (52). The two most important conditions are Clonorchiasis and opisthorchiasis. The flukes inhabit the bile ducts and are particularly prevalent in Southeast Asia where there is a tradition of eating raw fish.

One way to reduce health risks is, if there is sufficient space available, to rear the fish in 'unpolluted' water and to feed them on vegetable material e.g. duck weed which is grown in separate ponds which are fertilised with human or animal waste. The rationale behind this approach is the rapid reduction (up to 10,000-fold within 30 hours) in coliform and bacteriophage numbers that occurs in 'septage-fed' ponds (49). Delaying harvest by several days or weeks will clearly reduce the health risks to the 'harvesters' and will result in the fish consuming forage which should be relatively unpolluted.

Mycotoxins

The topic of mycotoxins in animal feed has been reviewed recently (53). The ill effects to both livestock and humans resulting from the consumption of mouldy grains has been recognised for centuries However, the importance of mycotoxins was only recognised formally during the 1960s when aflatoxin was shown to be responsible for a fatal condition of poultry termed 'Turkey X' disease. Since that time a considerable literature has accumulated and mycotoxicosis is recognised as a world wide problem. Mycotoxins can have a serious economic impact by causing losses in farm animals or giving rise to difficulties in their management, or by rendering a commodity unacceptable in national and international trade (54). Mycotoxins may be present in forage grasses and in harvested vegetable materials (53). Given sufficient resources it should be possible to control mycotoxin formation during storage although this may not be so easy in developing countries. The elaboration of mycotoxins in the field is more difficult to control and may require radical changes in agricultural practice in order to achieve this objective (54).

Cereals and other vegetable feed components are rarely harvested in a way which minimises microbial contamination. The low water activity of many of these materials restricts the growth of many microorganisms, however, although moulds may grow at relatively low levels (<0.80) and are responsible for the spoilage of many millions of tonnes of animal feed each year. In addition, moulds may render feed harmful by the elaboration of mycotoxins, of which many are now recognised including aflatoxin (*Aspergillus niger*), citrinin (*Penicillium citrinum, P. viridicatum*), fumonisin (*Fusarium* spp.*)*, ochratoxin (*A. Ochraceus, P. viridicatum*) and vomitoxin (*Fusarium* spp.). The mycology and toxicology of five agriculturally important mycotoxins, namely aflatoxin, deoxynivalenol, fumonisin, ochratoxin and zearalenone has been evaluated (55) while the literature concerning mycotoxins in ruminant rations and the role of mycotoxins in disease of animals including man has been reviewed (56).

Mycotoxins may be elaborated both before and after harvest (54) and the factors that affect fungal ecology before and after harvest have been discussed (57). Before-harvest fungicides and resistant plant varieties can be utilised to combat spoilage but a knowledge of the effects of weather on fungal ecology may assist the timing of remedial measures. Mould growth after harvest, can be controlled by the addition of propionic acid or other weak acids (57) but there is no effective way of destroying mycotoxins once formed, although treatment with ammonia at elevated temperatures and pressure (58), and the addition of non-nutritive sorbents to the diet which sequester the toxins, have been reported (59).

Humans are probably at greatest risk from mycotoxins following the consumption of cereals, legumes, pulses and vegetables. However, toxins ingested by animals can also pose a risk since they may be present in meat, milk or eggs (60) and may be found in sausages following production and aging (61).

The world-wide problem of mycotoxicosis is reflected by the fact that over 60 countries have either legislation or proposed legislation for the control of mycotoxins in both animal feed and human food (62). However, there is no consistent rationale for setting limits or for enforcement of control measures. Several certified reference materials for mycotoxins have been prepared by the European Commission's Community Bureau for References.

Pesticides and agricultural and industrial chemicals

Volumes 2A and 2B of the Codex Alimentarius provide a classification of foods and animal feeds and list Maximum Residue Limits (MRLs) for pesticides. Several definitions are provided including extraneous residue limit, good agricultural practice in the use of pesticides, maximum residue limit, pesticide and pesticide residue.

FAO has also published an international code of conduct on the distribution and use of pesticides (63). The document contains a section on reducing health hazards. There are no specific references to animal feed.

Drug residues

Volume 3 of the Codex Alimentarius contains, *inter alia*, MRLs for 15 veterinary drugs; an international code of practice for the control of the use of veterinary drugs; guidelines for the establishment of a regulatory programme for the control of veterinary drug residues in foods; and a comprehensive set of definitions including acceptable daily intake, bioavailable residue, bound residue, marker residue, maximum residue limit for veterinary drugs, non-extractable residue, veterinary drug and withdrawal and withholding time.

Heavy metals

Pollution of pasture and other farm land with heavy metals may occur either following aerial spread from smelting plants and other types of industrial processes, or the spreading of human and animal waste (sewage sludge, manure and slurry) on to land (39). In both circumstances, the metals may be ingested by animals grazing the pasture or consuming conserved forages and these metals may subsequently pass to the human population *via* meat or other products of animal origin. The uptake of heavy metals by plants is a complex matter and is influenced by the soil type (64). In addition, waste leather, which may contain chromium, may be inadvertently incorporated into animal feed.

Clearly, compounded animal feed should not contain excessive amounts of potentially toxic heavy metals. These metals and other elements can be divided into those that are highly toxic, toxic, moderately toxic and slightly toxic (see Table 2 above).

Radionuclides

The accidental emission of radioactive material from nuclear power stations is a well recognised problem and has very serious implications for the human population and for animals, both domesticated and wild. The Chernobyl disaster is the most serious of its kind to have occurred. A total of over 200 titles containing the word Chernobyl have been added to the Science Citation Index data base since 1981 and many of these are concerned with the contamination of pasture with radionuclides and their ingestion by farm livestock.

The prediction of the effects of a severe accident at a nuclear installation is a matter of great importance; the modelling of disasters of this kind, using computers has been discussed in recent publications (65,66). It may be necessary to feed animals with feed contaminated with radionuclides in the absence of non-contaminated products. The use of chemicals which either dilute the radionuclides with stable isotopes or analogous stable elements, or act as natural or artificial binding agents has been reviewed (67).

Conclusions

Animal feed or forage may be the source of a limited number of infections or intoxications for farm animals that could in theory lead to human illness. In practice, however, the risk to human health in some cases is negligible or non-existent e.g. *B. anthracis*, *Cl. botulinum* toxin, *Listeria monocytogenes* (except in the case of aborted foetuses and their membranes) and *M. bovis*. On the other hand, for the salmonellas the risk is real, and indeed there have been notable epidemics of human food-borne salmonellosis in which the original source of the serotype has been presumed to be animal feed (e.g. serotypes Agona and Virchow). The position concerning the BSE agent remains to be resolved although recent evidence (14) suggests that humans can become infected with the bovine prion. The extent of the risk has yet to be quantified accurately but it will probably be very small since, to date, there have only been a small number of cases of the NV-CJD in young people in the UK. The stringent control of SBM in the UK has removed the risk of these tissues entering the food chain.

Mycotoxins in animal feed can result in foods of animal origin containing these agents. This risk is well recognised but at present it has not been quantified accurately and in many instances may be of theoretical rather than of practical importance.

Pesticides, agricultural and industrial chemicals, heavy metals and radionuclides may pollute animal feed and forages. The methods for controlling pollution from these sources are well understood, from a technical point of view, although their implementation can prove difficult in many countries.

REFERENCES

1. HMSO. 1992. The Report of the Expert Group on Animal Feedingstuffs to the Minister of Agriculture, Fisheries and Food, the Secretary of State for Health and the Secretaries of State for Wales, Scotland and Northern Ireland. London, Her Majesty's Stationery Office.

2. FAO/WHO. 1995. Application of Risk Analysis to Food Standards Issues. Report of a Joint FAO/WHO Consultation. Geneva, WHO.

3. Hinton, M. & Bale, M.J. 1990. Animal pathogens in feed. *In* J. Wiseman & D.J.A. Cole eds, *Feedstuff Evaluation*, p 429-444. London, Butterworths.

4. Hinton, M. & Mead, G.C. 1990. The Control of feed-borne bacterial and viral pathogens in farm animals. *In* W. Haresign & D.J.A. Cole eds, *Recent Advances in Animal Nutrition 1990*, p 31-46. London, Butterworths.

5. Hinton, M. 1993. Spoilage and pathogenic microorganisms in animal feed. *International Biodeterioration and Biodegredation*, 32: 67-74.

6. Bradley, R. 1994. Bovine spongiform encephalopathy epidemiology: a brief review. *Livestock Production Science,* 38: 5-16.

7. Taylor, K.C. 1994. Bovine spongiform encephalopathy control in Great Britain. *Livestock Production Science,* 38: 17-21.

8. Anderson, R.M., Donnelly, C.A., Ferguson, N.M., Woolhouse, M.E.J., Watt, C.T., Udy, H.J., Mawhinney, S., Dunstan, S.P., Southwood, T.R.E., Wilesmith, J.W., Ryan, J.B.M., Hoinville, L.J., Hillerton, J.E., Austin, A.R. & Wells, G.A.H. 1996. Transmission dynamics and epidemiology of BSE in British cattle. *Nature,* 382: 779-788.

9. Bisping, W. 1993. Salmonella in feedingstuffs. *Deutsche Tierärtzliche Wöchenschrift,* 100: 262-263.

10. Bata, A., Glavits, R., Vanyi, A & Salyi, G. 1996. More important mycotoxicosis of poultry. *Magyar Allatorvosok Lapja,* 51: 395-408.

11. Dragon, D.C. & Rennie, R.P. 1995. The ecology of anthrax spores - tough but not invincible. *Canadian Veterinary Journal,* 35: 295-301.

12. WHO. 1996. Report of a WHO consultation on public health issues related to human and animal transmissible spongiform encephalopathies. Geneva, WHO.

13. FAO. 1993. Bovine Spongiform Encephalopathy by R.H. Kimberlin. *Animal Production and Health Paper No. 109.* Rome.

14. Collinge, J., Sidle, K.C.L., Meads, J., Ironside, J. & Hill, A.F. 1996. Molecular analysis of prion strain variation and the aetiology of 'new variant' CJD. *Nature,* 383: 666-667, 685-690.

15. Chambaud, L., Peters, P.W.J. & Merkel, B.C. 1996. Creutzfeldt-Jakob disease: results of an inquiry in the fifteen Member States of the European Union. *Eurosurveillance,* 1: 42-45.

16. Wyatt, J.M., Pearson, G.R. Smerdon, T., Gruffydd-Jones, T.J. & Wells, G.A.H. 1990. Spongiform encephalopathy in a cat. *Veterinary Record,* 126: 513.

17. Wyatt, J.M., Pearson, G.R. Smerdon, T., Gruffydd-Jones, T.J., Wells, G.A.H. & Wilesmith, J.W. 1991. Naturally occurring scrapie-like spongiform encephalopathy in five domestic cats. *Veterinary Record,* 129: 233-236.

18. Mohammed, M.D. & Hinton, M. 1993. The growth of salmonellas in hydrated animal feed supplemented with antibiotics. *Letters in Applied Microbiology,* 16: 7-9.

19. Berends, B.R., Urlings, H.A.P., Snijders, J.M.A. & Van Knapen, F. 1996. Identification and quantification of risk factors in animal management and transport regarding *Salmonella* spp. in pigs. *International Journal of Food Microbiology,* 30: 37-53.

20. Leeson, S. & Marcotte, M. 1993. Irradiation of poultry feed 1. Microbial status and bird response. *World Poultry Science Journal,* 49: 19-33.

21. Richter, A., Loscher, W & Witte, W. 1996. Feed additives with antimicrobial effects. *Praktische Tierarzt,* 77: 603.

22. Matthes, S. 1985. Influence of antimicrobial agents on the ecology of the gut and *Salmonella* shedding. *In* R. Helmuth & E. Bulling eds, *Criteria and methods for the microbial evaluation of growth promoters in animal feeds*, p 104-125. Berlin, BGA.

23. Barrow, P.A., Smith, H.W. & Tucker, J.F. 1984. The effect of feeding diets containing avoparcin on the excretion of salmonellas experimentally infected with natural sources of salmonella organisms. *Journal of Hygiene,* 93: 97-99.

24. Hinton, M., Al-Chalaby, Z.A.M. and Linton, A.H. 1986. The influence of dietary protein and antimicrobial food additives on salmonella carriage by broiler chicks. *Veterinary Record*, 119: 495-500.

25. Hinton, M.H., Allen, V.M. & Wray, C. 1992. The influence of growth promoting antibiotics on the colonization of the caecum of young chicks following consumption of feed artificially contaminated with salmonellas. *In* M.H. Hinton & R.W.A.W. Mulder eds, *Prevention and control of potentially pathogenic microorganisms in poultry and poultry meat processing. 7. The role of antibiotics in the control of pathogens*, p 69-75. Beekbergen, Agricultural Research Department (DLO-NL).

26. Blewett, D.A. 1984. The epidemiology of ovine toxoplasmosis. *Veterinary Annual,* 25: 120-124.

27. Radostits, O.M., Blood, D.C. & Gay, C.C. 1994. *Veterinary Medicine*, 8th Ed. London, Ballière Tindall.

28. Nesbakken, T. & Skjerve, E. 1996. Interruption of microbial cycles in farm animals from farm to table. *Meat Science,* 43: S47-S57.

29. Shih, J.C.H. 1993. Recent development in poultry waste digestion and feather utilization. *Poultry Science,* 72: 1617-1620.

30. Polprasert, C., Yang, P.Y., Kongsricharoern, N. & Kanjanaprapin, W. 1994. Productive utilization of pig farm wastes. A case study for developing countries. *Resources Conservation and Recycling,* 11: 1-4.

31. Williams, T.O., Powell, J.M. & Fernandezrivera, S. 1995. Soil fertility maintenance and food crop production in semi-arid West Africa. Is reliance on manure a sustainable strategy. *Outlook on Agriculture*, 24: 43-47.

32. Husted, S. 1994. Seasonal variation in methane emission from stored slurry and solid manures. *Journal of Environmental Quality,* 23: 585-592.

33. Steed, J. & Hashimoto, A.G. 1994. Methane emissions from typical manure management systems. *Bioresource Technology,* 50: 123-130.

34. Sutton, M.A., Place, C.J., Eager, M., Fowler, D. & Smith, R.I. 1995. Assessment of the magnitude of ammonia emissions in the United Kingdom. *Atmospheric Environment,* 29: 1393-1411.

35. Dong, H.M., Erda, L., Rao, M.J. & Yang, Q.C. 1996. An estimation of methane emissions from agricultural activities in China. *Ambio,* 25: 292-296.

36. Moore, P.A., Daniel, T.C., Edwards, D.R. & Miller, D.M. 1996. Evaluation of chemical amendments to reduce ammonia volatilization from poultry litter. *Poultry Science,* 75: 315-320.

37. Somner, S.G. & Hutchings, N. 1995. Techniques and strategies for the reduction of ammonia emission from agriculture. *Water and Soil Pollution,* 85: 237-248.

38. Goss, M.J. & Goorahoo, D. 1995. Nitrate contamination of groundwater. Measurement and prediction. *Fertilizer Research,* 42: 1-3.

39. Watson, D.C. 1985. Potential risks to human and animal health from land disposal of sewage sludge. *Journal of Applied Bacteriology,* 59, Symposium Supplement 95S-103S.

40. Strauch, D. 1987. *Animal Production and Environmental Health.* Amsterdam, Elsevier Science.

41. Vandenberg, J.J. 1993. Effects of sewage sludge disposal. *Land Degradation and Rehabilitation,* 4: 407-413.

42. Engstand, L. 1995. *Mycobacterium paratuberculosis* and Crohn's disease. *Scandinavian Journal of Infectious Diseases,* S98: 27-29.

43. Chiodini, R.J. & Rossiter, C.A. 1996. Paratuberculosis - a potential zoonosis. *Veterinary Clinics of North America - Food Animal Practice,* 12: 457.

44. Walton, J.R. & White, E.G. (Eds) 1981. *Communicable diseases resulting from storage handling, transport and landspreading of manures.* Luxembourg, Office for Official Publications of the European Communities.

45. Jones, P.W. 1980. Disease hazards associated with slurry disposal. *British Veterinary Journal,* 136: 529-542.

46. Jones, P.W. 1980. Health hazards associated with the handling of animal wastes. *Veterinary Record,* 106: 4-7.

47. Goodger, W.J., Collins, M.T., Nordlund, K.V., Eiselle, C., Pelletier, J., Thomas, C.B. & Sockett, D.C. 1996. Epidemiologic study of on-farm management practices associated

with the prevalence of *Mycobacterium paratuberculosis* infections in dairy cattle. *Journal of the American Veterinary Medical Association,* 208: 1877-1881.

48. Edwards, P. 1992. *Reuse of Human Wastes in Aquaculture.* Washington, D.C., The World Bank.

49. Edwards, P. 1996. Wastewater-fed aquaculture systems: status and prospects. *The ICLARM Quarterly,* January 1996: 33-35.

50. Feacham, R.G., Bradley, D.J., Garelick, H. & Mara, D.D. 1983. *Sanitation and Disease: Health Aspects of Excreta and Wastewater Management.* Chichester, UK John Wiley.

51. Association of American Feed Control Officials. 1996. Official Publication. p 230.

52. Ko, R.C. 1995. Fish-borne parasitic zoonoses. *In* P.T.K. Woo ed, *Fish Diseases and Disorders. Volume 1. Protozoan and Metazoan Infections,* p 631-671. Wallingford, UK, CAB International.

53. Meronuck, R. & Cincibido, V. 1996. Mycotoxins in feed. *Feedingstuffs,* 68: 139-145.

54. Moss, M.O. 1991. Economic importance of mycotoxins - recent incidence. *International Biodeterioration,* 27: 195-204.

55. Miller, J.D. 1993. Fungi and mycotoxins in grain - implications for stored product research. *Journal of Stored Product Research,* 31: 1-16.

56. Krogh, P. 1989. The role of mycotoxins in diseases of animals and man. *Journal of Applied Bacteriology,* 67, Symposium Supplement 99S-104S.

57. Lacey. J. 1989. Pre- and post-harvest ecology of fungi causing spoilage of foods and other stored products. *Journal of Applied Bacteriology,* 67, Symposium Supplement 11S-25S.

58. Park, D.L. 1993. Perspectives of mycotoxin decontamination procedures. *Food Additives and Contaminants*, 10: 49-60.

59. Ramos, A.J., Finkgremmels, J.& Hernandez, E. 1996. Prevention of toxic effects of mycotoxins by means of nonnutritive adsorbant compounds. *Journal of Food Protection,* 59: 631-641.

60. Park, D.L. & Liang, B.L. 1993. Perspectives of aflatoxin control of human food and animal feed. *Trends in Food Science and Technology,* 4: 334-342.

61. Luskey, K., Tesch, D. & Gobel, R. 1995. Effect of natural and crystalline ochratoxin A in pigs after a feeding period of 28 days, and behaviour of the toxin residues in body fluids, organs and meat products. *Archiv für Lebemittelhygiene,* 46: 45-47.

62. Vanegmond, H.P. 1995. Mycotoxins - regulations, quality assurance and reference materials. *Food Additives and Contaminants,* 12: 321-330.

63. FAO. 1990. *International Code of Conduct on the Distribution and Use of Pesticides.* Rome.

64. Miller, R.W., Azzari, A.S. & Gardiner, D.T. 1995. Heavy metals in crops as affected by soil types and sewage sludge rates. *Communications in Soil Science and Plant Analysis,* 26: 5-6.

65. Helton, J.C., Johnson, J.D., Rollstin, J.A., Shiver, A.W. & Sprung, J.L. 1995. Uncertainty and sensitivity analysis of food pathway results with the MACCS reactor accident consequence model. *Reliability Engineering and System Safety,* 49: 109-144.

66. Helton, J.C., Johnson, J.D., Rollstin, J.A., Shiver, A.W. & Sprung, J.L. 1995. Uncertainty and sensitivity analysis of chronic exposure results with the MACCS reactor accident consequence model. *Reliability Engineering and System Safety,* 50: 137-177.

67. Voight, G. 1993. Chemical methods to reduce the radioactive contamination of animals and their products in agricultural ecosystems. *Science of the Total Environment,* 137: 205-225.

CODEX ALIMENTARIUS COMMISSION

Standards, Guidelines and Other Recommendations
Related to the Quality and Safety of Feeds and Foods

The Codex Alimentarius Commission is responsible for implementing the Joint FAO/WHO Food Standards Programme. The name *Codex Alimentarius* is taken from Latin and translates literally as "food code" or "food law". It was founded in response to the world-wide recognition of the importance of international trade, the need to facilitate such trade, while at the same time ensuring the quality and safety of food for the world consumer.

The Commission has, as its primary objective, the protection of the health of the consumers, the assurance of fair practices in the food trade and the co-ordination of all food standards work. Formulation of food standards, guidelines and recommendations is the work of the Commission. With the adoption of the World Trade Organization's Agreement on the Application of Sanitary and Phytosanitary Measures and the Agreement on Technical Barriers to Trade, a new emphasis and dimension have been placed on Codex standards.

A number of Codex standards, guidelines and recommendations already include provisions which relate to the quality and safety of animal feeds. These include:

1. **Codex General Standard for Contaminants and Toxins in Food (Codex Stan 193-1995 - Volume 1A, Section 6.1)**

    This standard contains the main principles and procedures which are used and recommended by the Codex Alimentarius in dealing with contaminants and toxins in food and feeds and lists the maximum levels of contaminants and natural toxicants in food and feeds which are recommended by the Commission to be applied to commodities moving in international trade.

2. **List of Codex Maximum Residue Limits (MRLs) for Pesticides and Codex Extraneous Maximum Residue Limits (EMRLs) (General Text) (Volume 2A and MRLs, Volume 2B)**

    This text provides the basis for establishing the Codex MRLs and EMRLs, the consideration given to human daily intake and explanatory notes on how to view and interpret the data, describing the meaning of various symbols used. A list of the commodities and Codex MRLs/EMRLs is provided.

3. **List of Codex Maximum Residue Limits (MRLs) for Veterinary Drugs (Volume 3-1994)**

    The Codex MRLs for Veterinary Drugs are consistent with the recommendations of the Joint FAO/WHO Expert Committee on Food Additives (JECFA), a body of independent scientists who serve in their individual capacity as experts and evaluate veterinary drugs to establish safe levels of intake and develop maximum residue levels when veterinary drugs are used in accordance with good veterinary practices.

4. **Recommended International Code of Practice for Control of the Use of Veterinary Drugs (CAC/RCP 38-1993, Volume 3-1994)**

This code sets out guidelines on the prescriptions, application, distribution and control of drugs used for treating animals, processing animal health and improving animal production. It includes Good Practices in the Use of Veterinary Drugs (GPVD), including premixes for the manufacture of medicated feedingstuffs.

5. **Codex Standards for Processed Meat and Poultry Products (Part 1, Volume 10-1994)**

A number of food standards have been elaborated by Codex with quality and safety requirements. They include standards for Corn Beef, Luncheon Meat, Cooked Cured Ham, Cooked Cured Pork Shoulder and Cooked Cured Chopped Meat.

6. **Recommended International Code for Ante-Mortem and Post-Mortem Inspection of Slaughter Animals and for Ante-Mortem and Post-Mortem Judgment for Slaughter Animals and Meat (CAC/RCP 41-1993, Part 3, Volume 10-1994)**

This Code, together with the Code of Hygienic Practice for Fresh Meat (CAC/RCP 11-1976, Rev. 1 (1993), describes requirements necessary to achieve acceptable levels of safety and wholesomeness for fresh meat from slaughtered animals throughout the food chain starting from the farm of origin.

7. **Other Codes of Practices and Guidelines for Processed Meat and Poultry Products (Part 2 and Part 3, Volume 10-1994)**

Codex has elaborated recommendations and guidelines related to hygienic protection for Processed Meat and Poultry Products; Poultry Processing; Production, Storage and composition of Mechanically Separated Meat and Poultry Meat intended for further Processing; Hygienic Practice for Fresh Meat and Game.

These standards, recommendations and guidelines relate to the quality and safety of the animal origin products resulting from methods and procedures utilized in, and including feeding of, production animals.

FAO TECHNICAL PAPERS

FAO FOOD AND NUTRITION PAPERS

Availability: June 1998

Ar	– Arabic	Multil	– Multilingual
C	– Chinese	*	Out of print
E	– English	**	In preparation
F	– French		
P	– Portuguese		
S	– Spanish		

The FAO Technical Papers are available through the authorized FAO Sales Agents or directly from Sales and Marketing Group, FAO, Viale delle Terme di Caracalla, 00100 Rome, Italy.